MW00711693

TABLE OF CONTENTS (continued)

UGLY'S ™

Electrical Safety and NFPA 70E®

For Systems Rated at 600 Volts and Below

by H. Brooke Stauffer

Printed in the U.S.A.

A note from the author . . .

Ugly's Electrical Safety and NFPA 70E is designed to be used as a quick on-the-job reference covering the key requirements for electrical safety in an easy-to-read format.

Ugly's Electrical Safety and NFPA 70E is not intended to be a substitute for *NFPA 70E Standard for Electrical Safety in the Workplace* or *NFPA 70E Handbook for Electrical Safety in the Workplace*.

We salute the National Fire Protection Association for their dedication to the protection of lives and property from fire and electrical hazards through the sponsorship of the *NFPA 70E Standard for Electrical Safety in the Workplace* and *NFPA 70E Handbook for Electrical Safety in the Workplace*.

NFPA 70E Standard for Electrical Safety in the Workplace and *NFPA 70E Handbook for Electrical Safety in the Workplace* are registered trademarks of the National Fire Protection Association, Inc., Quincy, MA.

NFPA®

JONES AND BARTLETT PUBLISHERS
Sudbury, Massachusetts
BOSTON   TORONTO   LONDON   SINGAPORE

TABLE OF CONTENTS

TABLE OF CONTENTS (continued)

TABLE OF CONTENTS (continued)

TABLE OF CONTENTS (continued)

 **INTRODUCTION**

The purpose of *Ugly's Electrical Safety and NFPA 70E* ® is to provide recommendations in an abbreviated form where the circuit voltage does not exceed 600 volts. This *Ugly's* guide is based on *NFPA 70E, Standard for Electrical Safety in the Workplace.* This publication is shorter than the actual standard and does not cover everything in it. This guide should be used *with NFPA 70E—not as a replacement for it.* For more complete information about electrical safety, see the following:

• *NFPA 70E—2009, Standard for Electrical Safety in the Workplace*

• *NFPA 70E: Handbook for Electrical Safety in the Workplace*

⬛ WHAT THIS GUIDE DOES <u>NOT</u> COVER

1. Above 600 volts

This guide applies to work on systems rated at 600 volts or below. It covers typical "inside" electrical work at the following voltages:

- 480-volt, 3-phase, 3-wire systems

- 277/480-volt, 3-phase, 4-wire systems

- 120/208-volt, 3-phase, 4-wire systems

- 120/240-volt, single-phase, 3-wire systems

Generally, this guide does not cover power systems operating at higher voltages, such as those at industrial plants or campus-wide power distribution systems, and we have included minimal coverage of these topics. For additional information on safe work practices at higher voltages, see the following:

- *NFPA 70E—2009, Standard for Electrical Safety in the Workplace*

- *NFPA 70E: Handbook for Electrical Safety in the Workplace*

WHAT THIS GUIDE DOES NOT COVER

2. Utility Systems

Utility systems used for generation, transmission, and distribution are outside the scope of *NFPA 70E*. For this reason, *Ugly's Electrical Safety and NFPA 70E* does not cover work in utility generating plants or sub-stations, or on transmission lines, distribution lines, or street lighting systems owned and operated by electric utilities.

3. Non-Electrical Hazards

This guide describes how to protect electrical workers from electric shock and arc-flash hazards. It does not cover protection from other hazardous conditions such as falls, use of ladders and scaffolds, and hazardous substances.

FOUR PROTECTIVE STRATEGIES

The *NFPA 70E* standard describes safe practices for workers who install or maintain electrical conductors and equipment. The standard embraces four different protective strategies to eliminate or minimize exposure to electrical hazards. Requirements associated with the following four strategies discuss the most protective strategy to the least protective.

1. **Turn off the power.**
   Work deenergized, whenever possible. The standard recognizes that some work tasks, such as measuring voltage, require the circuit to be energized. When working within the limited approach boundary and the arc-flash protection boundary of exposed conductors and parts that are or might be energized, workers (or their supervisor) must use strategies 2–4.

2. **Use an energized work permit.**
   Have the customer sign an Energized Electrical Work Permit (EEWP). Additionally, if the worker is employed by the facility, the facility safety program should result in an EEWP.

3. **Plan the work.**
   Have a written plan for performing the work safely.

4. **Use PPE (Personal Protective Equipment).**
   This includes FR (flame resistant) clothing, insulated tools, face shields, and flash suits.

⚡ QUALIFIED PERSONS (ELECTRICAL WORKERS)

Both the *National Electrical Code (NEC)*® and *NFPA 70E* define *qualified person* as "one who has skills and knowledge related to the construction and operation of the electrical equipment and installations and has received safety training to recognize and avoid the hazards involved." Only people who have received this training are permitted to do work within the limited approach boundary **(see Figure 1)**.

FIGURE 1   Shock approach boundaries.

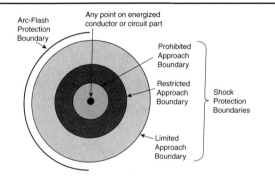

Reprinted with permission from NFPA 70E®, *Handbook for Electrical Safety in the Workplace*, Copyright © 2009, National Fire Protection Association. This reprinted material is not the complete and official position of the NFPA on the referenced subject, which is represented only by the standard in its entirety.

QUALIFIED PERSONS (ELECTRICAL WORKERS)

Qualified persons or workers who install and maintain electrical systems are electricians, engineers, technicians, and building maintenance personnel who have received electrical training. Only qualified persons are permitted to perform maintenance on electrical equipment, circuits, and installations.

A person is not necessarily qualified if they are licensed. If a person is trained and qualified on Vendor A switchgear, panelboards, and so on, they may not be qualified on Vendor B switchgear, panelboards, and so on, because they may operate or disengage differently.

Protecting Others (That Is, Unqualified Persons)

NFPA 70E defines rules for protecting untrained workers from electrical hazards. Generally, people such as office workers, teachers, students, retail employees, customers, visitors, healthcare workers, cleaning crews, and workers in non-electrical trades such as plumbers, carpenters, and painters are considered to be unqualified persons.

NFPA 70E Requirements

Warning and Guarding

* Warning signs and barriers keep unqualified persons away from places where an electrical hazard exists due to ongoing electrical work.

Maintenance and Housekeeping

* Adequate maintenance and housekeeping make it possible to work safely on equipment by keeping the equipment in safe operating condition.

* Good maintenance includes technical factors, such as ensuring that all electrical boxes have covers and all electrical equipment is adequately grounded.

QUALIFIED PERSONS (ELECTRICAL WORKERS)

- Good housekeeping includes non-technical factors, such as not using electrical rooms and closets for storage purposes.

- Workspaces around electrical equipment must be kept clear so that qualified persons can do maintenance and repair work safely.

- Emergency access to switches and circuit breakers is essential. Space in front of electrical disconnecting means must be kept clear.

ESTABLISHING AN ELECTRICALLY SAFE WORK CONDITION

Turning off the power, verifying that it is off, and ensuring that it stays off while work is being performed is called *establishing an electrically safe work condition*. Six steps are involved in this process:

1. Identify power sources.

2. Disconnect power sources.

3. Verify that power is disconnected.

4. Use lockout–tagout.

5. Verify that power is off with test instruments.

6. Discharge stored electrical energy (and install safety grounds under certain conditions).

Lockout–Tagout

Lockout–tagout is an important part of the overall process. However, lockout–tagout is only one of the six steps in *establishing an electrically safe work condition*.

Warnings

- Only qualified persons are permitted to establish an electrically safe work condition.

- The process of establishing an electrically safe work condition is inherently hazardous. It requires contact (such as voltage testing) with exposed electrical conductors or circuit parts that might be energized. Electrical workers must wear appropriate PPE when performing some of the steps.

- Electrical conductors and equipment are considered energized until the process of *establishing an electrically safe work condition* is complete.

ESTABLISHING AN ELECTRICALLY SAFE WORK CONDITION

Six-Step Procedure for Establishing an Electrically Safe Work Condition

1. **Identify the power source.**
   - Determine all possible sources of electric supply to the equipment to be worked on.
   - Check electrical plans, one-line diagrams, panelboard schedules, and identifying signs and tags on electrical equipment.
   - Most electrical equipment has a single source of supply, but sometimes equipment might have multiple sources. The multiple sources might include emergency generators, interactive power sources such as photovoltaic or fuel cell systems, and dual utility feeds for major industrial facilities.
   - Sometimes "illegal" circuits are installed that do not comply with *NEC* rules. These can create backfeed hazards after workers have disconnected all the electrical power sources they know about. An example is a 480v/120v lighting transformer. If the 480v is deenergized for working and temporary power is connected to the 120v panel, it must be verified that the 120v power cannot back-feed to the 480v system because 480v breakers were not opened and locked out.

2. **Disconnect power sources.**
   - After properly interrupting the load current, open the disconnecting means for each source.
   - Most circuit breakers, safety switches, and other disconnecting means are capable of interrupting the load current they carry.
   - When the rating of a disconnect is not sufficient to interrupt the load current, the load must be removed by another operation before the handle is operated.

ESTABLISHING AN ELECTRICALLY SAFE WORK CONDITION

- Fuses are not considered disconnecting means, so a circuit cannot be deenergized merely by removing one or more plug or cartridge fuses. However, a pullout block (range fuse block) or safety switch with fuses is considered a disconnect. Operating the switch or pulling out the fuse block disconnects all ungrounded (phase) conductors downstream of the circuit.
- On most premise wiring systems, only the ungrounded (phase) conductors are disconnected. The grounded (neutral) conductors normally are not interrupted intentionally.
- Attachment plugs for electrical appliances such as cooking and laundry equipment can be used as disconnects.

3. **Verify that power conductors are opened.**
   - Wherever possible, the worker must visually verify that all blades of the disconnecting means are fully open, or that drawout-type circuit breakers are racked out to their fully disconnected position.
   - Disconnecting means sometimes malfunction and fail to open all phase conductors when the handle is operated. After operating the disconnect's handle, a qualified person should open the equipment door or cover and look to see that a physical opening (air gap) exists in each blade of the disconnect.

      ⚠ *WARNING: Electrical workers must wear both shock and arc-flash PPE when performing this operation.*

   - Sometimes it is impossible to visually verify the existence of an air gap. (For example, no air gap will be visible in a molded-case circuit breaker.) In these cases, the qualified person must test for the presence of voltage to verify that the circuit has actually been disconnected. Using a voltmeter rated for that voltage level, he or she tests for voltage between all phase conductors and between each phase conductor and ground (phase-to-phase and phase-to-ground).

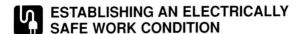 # ESTABLISHING AN ELECTRICALLY SAFE WORK CONDITION

> ⚠ *WARNING: Electrical workers must wear both shock and arc-flash PPE when performing this operation.*

4. **Lockout–tagout.**
   - Apply lockout–tagout devices in accordance with the employer's written electrical safety program. Normally these devices are padlocks to keep the disconnecting means open, and tags that identify the person(s) responsible for applying and removing the locks **(see Figure 2)**.

FIGURE 2   A lockout station. (Photo courtesy of Grainger)

5. **Verify that no voltage exists.**
   - Test for the presence of voltage.
   - The functionality of the voltmeter must be verified on a known source both before and after using it to test for the presence of voltage.
   - The worker should use a voltmeter or other tester rated Category III or IV to test conductors and equipment operating at up to 480 volts.
   - Testers rated Category II can be used on single-phase, 120-volt circuits.

   ⚠ *WARNING: Electrical workers must wear both shock and arc-flash PPE when performing this operation. Also, see the "Test Meter Safety Ratings."*

6. **Discharge stored electrical energy (and install safety grounds under certain conditions).**
   - The worker must discharge sources of stored energy such as capacitors used for power factor correction and motor starting.

   ⚠ *WARNING: Electrical workers must wear both shock and arc-flash PPE when performing this operation.*

How to Read a Warning Label

Equipment shall be field marked with a label containing the available incident energy or required level of PPE **(see Figure 3)**.

ESTABLISHING AN ELECTRICALLY SAFE WORK CONDITION

FIGURE 3 Warning label. (Reprinted with permission from Littelfuse®, www.littelfuse.com, 1-800-TEC-FUSE)

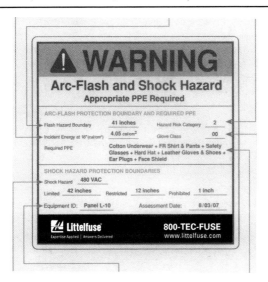

Electrically Safe Work Condition Established

Once the preceding steps have been completed, electrical energy has been removed from all conductors and equipment and cannot reappear unexpectedly. Under these circumstances, arc-rated or voltage-rated (insulated) electrical PPE is not needed, and unqualified persons can perform work such as cleaning and painting on or near electrical equipment. However, only electrically qualified persons should perform technical work within the scope of the *National Electrical Code*. This is true whether or not the electrical system is energized.

13

⚡ TEST METER SAFETY RATINGS

Underwriters Laboratories, Inc. (UL) has four safety categories for test and measurement equipment. They are CAT I, CAT II, CAT III, and CAT IV, with CAT IV offering the highest level of protection.

- Meters are subject to transient voltage that can result from a lightning discharge, switching spike, or other reasons.

- Voltmeters are assigned a transient rating based on the ability of the meter to continue working after experiencing a voltage spike.

- A worker must never connect any meter to electrical conductors or equipment with voltage or current higher than the rating of the meter itself.

- Transient over-voltages (spikes) caused by nearby lightning strikes, utility switching, motor starting, and capacitor switching can damage the electronic circuitry inside the meter, and can even cause it to explode while being used.

- Listed digital multimeters (DMMs) have internal fuses to protect the test instrument (and the person using it), but the worker must make sure that the meter is properly rated for the application before taking a reading **(see Table 1)**.

TABLE 1   Meter Safety—UL Category Ratings

CAT I Isolated equipment	Equipment plugged into receptacle outlets and not directly connected (hard wired) to the building supply system.
CAT II Directly-connected circuits	Single-phase circuits supplied from panelboards. Equipment and appliances supplied by single-phase branch circuits.
CAT III Inside building circuits	Service equipment, panelboards, motor control centers. Three-phase feeders and branch circuits. Single-phase branch circuits supplied directly from the service equipment.

QUALIFIED PERSONS (ELECTRICAL WORKERS)

Both the *National Electrical Code (NEC)*® and *NFPA 70E* define *qualified person* as "one who has skills and knowledge related to the construction and operation of the electrical equipment and installations and has received safety training to recognize and avoid the hazards involved." Only people who have received this training are permitted to do work within the limited approach boundary **(see Figure 1)**.

FIGURE 1   Shock approach boundaries.

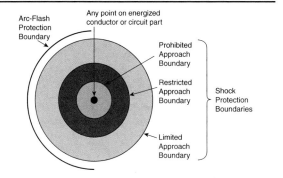

⬛ QUALIFIED PERSONS (ELECTRICAL WORKERS)

Qualified persons or workers who install and maintain electrical systems are electricians, engineers, technicians, and building maintenance personnel who have received electrical training. Only qualified persons are permitted to perform maintenance on electrical equipment, circuits, and installations.

A person is not necessarily qualified if they are licensed. If a person is trained and qualified on Vendor A switchgear, panelboards, and so on, they may not be qualified on Vendor B switchgear, panelboards, and so on, because they may operate or disengage differently.

Protecting Others (That Is, Unqualified Persons)

NFPA 70E defines rules for protecting untrained workers from electrical hazards. Generally, people such as office workers, teachers, students, retail employees, customers, visitors, healthcare workers, cleaning crews, and workers in non-electrical trades such as plumbers, carpenters, and painters are considered to be unqualified persons.

NFPA 70E Requirements

Warning and Guarding

- Warning signs and barriers keep unqualified persons away from places where an electrical hazard exists due to ongoing electrical work.

Maintenance and Housekeeping

- Adequate maintenance and housekeeping make it possible to work safely on equipment by keeping the equipment in safe operating condition.

- Good maintenance includes technical factors, such as ensuring that all electrical boxes have covers and all electrical equipment is adequately grounded.

 **TEST METER SAFETY RATINGS**

TABLE 1   Meter Safety—continued

CAT IV Supply conductors Outdoor conductors	Service drop and service lateral conductors. Watt-hour meter. Line side of the main service disconnect. Outdoor feeders and branch circuits.

Location Is Everything

High-voltage transients can be caused by nearby lightning strikes, equipment within the facility, and normal utility switching operations. For this reason, choosing the right test meter for a particular measurement does not depend solely on circuit voltage or current, but on a combination of these factors and on the location of the item being tested in the premises wiring system **(see Figure 4)**.

- **CAT IV** meters can be used safely on outdoor conductors, the main lugs or main overcurrent (protection) device of service equipment, and watt-hour meters.

- **CAT III** meters can be used safely on power systems inside buildings and similar structures. These systems include panelboards, motor control centers, feeders, busways, motors, and hardwired luminaires (lighting fixtures).

- **CAT II** meters can be used safely on single-phase, receptacle-connected loads located more than 33 feet (10 meters) from a CAT III power source or more than 66 ft (20 meters) from a CAT IV source.

- **CAT I** meters are intended for use only on electronic equipment.

TEST METER SAFETY RATINGS

FIGURE 4 A meter typical of those used to test for voltage.
(Courtesy of Fluke. Reproduced with permission.)

Meter Marking and Accessories

Meters are marked by the manufacturer with their category ratings. They are often dual-rated, such as 1000V CAT II / 600V CAT III. If a meter does not have a marked category rating, it is a CAT I device. Meters without a category marking provided by the manufacturer should not be used.

These energy safety ratings apply also to test leads, clamp-on adapters, and any other electrical accessories used with the meter. The lowest rating on any of these is the "weak link" that determines the overall category rating for that measurement.

 **TEST METER SAFETY RATINGS**

Select the Proper Rating

Workers should select a meter with a rating that exceeds the anticipated application requirements for both the category and voltage. For
most industrial work, 600V CAT III is the minimum acceptable.

Hazardous (Classified) Locations

The four UL category ratings do not apply to meters used in explosive
atmospheres. Any meter used in a hazardous (classified) location as
defined by the *NEC* must be "intrinsically safe."

Voltmeters

Many different types of voltmeters and voltage detectors are available
for purchase **(see Figure 5 and Figure 6)**. In most instances, the
devices work well within the intended parameters. To avoid misapplication of the devices, workers must be familiar with the intended purpose
of the device.

- For a worker to be safe from an electrical incident, the most important thing that he or she must know is whether an electrical conductor is energized.

- If the electrical conductor is energized, the next most important bit of
  information is the level of the voltage that is present.

- Only a voltmeter can determine whether a conductor is energized.
  Although the voltage-detecting device is not worn or installed, as is
  the case with other PPE, the construction and integrity of the device
  are critical.

- Should a voltmeter fail while in direct contact with an exposed energized conductor, an arcing fault may result.

FIGURE 5   Voltmeter.

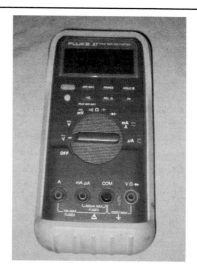

- Should a voltmeter fail to accurately sense the presence of voltage, electrocution may result. A worker can sense the presence of voltage by using an effective voltage detector. A voltmeter can also be used if it is tested on a known source, used to check an item in question, and then used on a known source again to make sure it is working.

- Many different types of voltmeters and voltage detectors are available for purchase.

- In most instances, the devices work well within the intended parameters. To avoid misapplication of the devices, workers must be familiar with the intended purpose of the device.

TEST METER SAFETY RATINGS

FIGURE 6   **Voltage detector.** (Courtesy of TEGAM)

The Home Environment

Voltmeters are available from many manufacturers and come in many different sizes and configurations. Some devices are intended for use by a hobbyist working on circuits with limited capacity. Usually, these meters are sold in home improvement and hobby stores.

- Home improvement stores stock voltmeters that are likely to be used on circuits in the home environment.

- In a home environment, the system capacity could be sufficient to sustain an arcing fault. Electrical circuits in the range of 120 volts to 240 volts are high enough to result in electrocution.

⬛ TEST METER SAFETY RATINGS

- Although the capacity of a residential system is much smaller than a commercial or industrial system, the integrity of a voltage-measuring device is equally important.

Selection and Use

Users should select a voltmeter that has a range commensurate with the expected circuit voltage.

- If the voltmeter is intended to verify the absence of voltage, users should select a device intended for direct contact.

- If the level of voltage is not known, users should select a meter that has auto-ranging capability, but users must ensure that the circuit voltage is lower than the maximum range of the meter.

When a voltage-detecting device is being used, electrical injuries occur for one or more of the following reasons:

- The voltmeter was misapplied or misused.

- The voltmeter was selected improperly.

- The indication was misunderstood.

- The leads came out of the meter and touched the grounded enclosure, resulting in a short circuit.

- The user's hand slipped off the end of the probe and contacted the energized conductor.

- An internal failure occurred, and the meter exploded.

TEST METER SAFETY RATINGS

When the probes from a voltmeter are in contact with energized conductors, the circuit under test experiences an additional circuit element. Current flows between one voltmeter probe and the other.

- The amount of current depends on the internal impedance of the voltmeter.

- Current is measured in volts on a graduated scale (either analog or digital). In solenoid-type devices, the amount of current flow exerts a known magnetic force on the solenoid.

- The solenoid movement is graduated in voltage.

- For the voltmeter to function effectively, the integrity of the current path through the measuring element is critical.

- The internal current path includes the probes, the plug in the case that accepts the probes, and internal components.

- In some cases, current flows through the switch used for changing scale. The switch easily could be set to the incorrect position, which would destroy the meter and could cause an arcing fault.

- Devices with low internal impedance tend to discharge induced or static voltage. High internal impedance devices may measure all voltage, including induced and static voltage.

- Workers must understand the relationship of internal impedance to measuring voltage and the internal impedance of the current path.

- Some voltage-detecting devices do not require direct contact with an exposed energized conductor. Noncontact voltage detecting devices sense the presence of an electrostatic or electromagnetic field. These devices look for inductive or capacitive coupling between the device and the conductor in question.

- Normally, noncontact voltage detectors provide an audible signal that a voltage is present.

- Some devices also provide a visual signal. Depending on how the noncontact voltage detector works and the physical construction of the conductor in question, a null point may exist along the conductor's linear direction. These devices provide an initial indication of the presence of voltage, but they should not be used to determine the absence of voltage.

Duty Cycle

A duty cycle is intended to prevent the device from overheating. Several injuries occur every year because a solenoid-type device overheats and explodes. Solenoid-type voltage-detecting devices are constructed by wrapping a small wire around a core to form the coil of a solenoid.

- Depending on the construction of the device, the manufacturer might assign a duty cycle to the device.

- The small wire usually has varnish for insulation. To avoid overheating the insulating varnish, a duty cycle is assigned to permit the coil to cool. A rule of thumb is that a solenoid voltage detector should be used for no more than 15 seconds without permitting the device to cool for at least another 15 seconds.

- Manufacturers define the duty cycle on the label attached to the equipment.

Although a voltmeter may be used to troubleshoot a circuit, the voltmeter is a safety device.

- A voltmeter provides crucial information for a worker to evaluate his or her exposure to an electrical hazard. The voltmeter is PPE in the same vein as safety glasses (spectacles).

- Electricians sometimes keep a voltmeter in their toolbox with screwdrivers, socket wrenches, pipe wrenches, and so on. Workers then tend to view their voltmeter as a hand tool that is used to troubleshoot a circuit, giving little thought to the fact that their lives depend on the integrity of their voltmeters.

⏻ TEST METER SAFETY RATINGS

- Some workers carry voltmeters that were purchased at a hobby store because those devices may be cheaper.

- Still other workers carry a shirt pocket version of a noncontact voltage detector.

- These approaches may prevent workers from accurately sensing the presence or absence of voltage in a circuit, which is the first step in preventing injury.

UL 1244 Requirements

The worker should purchase a voltmeter that complies with specifications relevant for the particular intended use of the voltmeter. The national consensus standard covering voltmeters is *UL 1244, Standard for Electrical and Electronic Measuring and Testing Equipment*. Requirements defined in this standard address issues that result in injuries associated with the construction of voltmeters that were known when the standard was promulgated.

For instance, *UL 1244* requires that:

- the banana plugs that connect the leads to the voltmeter be shielded to prevent contact with a grounded surface in case either of the plugs slips out of the receptacle.

The standard also requires that:

- probes contain a knurled section near the end to help prevent a worker's hand from slipping and contacting the energized conductor.

- the meter be designed such that the mode selector switch cannot be a part of the active circuit.

- adequately rated fuses, eliminating the risk of initiating an arcing fault from component failure, be present.

⟲ TEST METER SAFETY RATINGS

Only voltage-detecting devices that are evaluated and comply with
UL 1244 may display the UL label. Unless products are so marked, they
do not comply with the standard's requirements.

Electrical systems are becoming increasingly complex. The amount of
equipment that generates transient voltage spikes on a system is
increasing.

- A transient voltage spike results from a static discharge in a lightning
  strike.

- Transients also might be the result of switching inductive loads.

- Normally, voltage spikes are a few microseconds in duration but can
  involve many hundreds of amperes. In some areas of North America,
  lightning discharges are common, especially in the spring and sum-
  mer months.

- When a transient spike is in an electrical circuit, any equipment electrically connected to that circuit must be capable of handling the
  spike.

- Transients can destroy a voltmeter that happens to be in contact with
  the electrical conductor simultaneously.

- Transients contain less energy as the distance from the source of the
  transient increases.

Static Discharge Categories

*ANSI/ISA Standard S82.02.01, Electrical and Electronic Test,
Measuring, Controlling, and Related Equipment, General Requirements,*
and similar international standards establish a rating system for
voltmeters.

Voltmeters and other measuring instruments are assigned to categories. The categories differentiate the ability of the devices to handle
the energy in a transient condition (such as a spike in voltage caused

⚡ TEST METER SAFETY RATINGS

by lightning or switching). The categories are established by the location in the circuit between the source of electricity and the point where the device will be used.

- Voltmeters that can be used in the point of generation and transmission (where most energy is available) are Category (CAT) IV devices.

As the distance from the generator or transmission line to the point of use increases, the assigned category decreases.

- CAT III devices can be used safely with distribution level circuits such as motor control centers, load centers, and distribution panels.

- CAT II devices can be used safely with receptacles and utilization circuits. CAT I devices are intended for use with electronic equipment and circuits.

As the category decreases (from IV to I), the ability of the device to resist damage from transient over-voltage decreases. Therefore, CAT I devices are less likely to survive a transient over-voltage (spike).

In the international community, voltmeters are assigned to categories according to their ability to function in various environments where transient currents are expected. The international standard that covers transient categories for voltmeters is *IEC 61010, Safety Requirements for Electrical Equipment for Measurement, Control, and Laboratory Use.* The transient categories assigned by *IEC 61010* are CAT I, CAT II, CAT III, and CAT IV. These transient category ratings align with ratings established by the ANSI/ISA system. Voltmeter ratings based on *ANSI/ISA S82.02.01* are equivalent to the ratings based on *IEC 61010.*

Purchase

Voltmeters provide information that is critical to preventing injuries. Electricians and technicians often keep their voltmeters in the same toolbox where they keep socket sets, screwdrivers, pipe wrenches, and other hand tools. As the toolbox is bumped and dropped, the critical

TEST METER SAFETY RATINGS

voltmeter is subjected to the same physical forces as the metal tools. The voltmeter case or internal components may be damaged by the physical shock. Voltmeters that will be subjected to physical shocks should meet a specification that provides for protection from the physical abuse. The authors recommend that the purchase order for voltmeters require the device to have a UL label indicating compliance with *UL 1244*. The authors also recommend that voltmeters for use in an industrial setting be assigned a CAT IV rating, as determined by *ANSI S82.02.01* or *IEC 61010*.

Inspection

Users should inspect each voltmeter each day before use. The user must always ensure that the voltmeter is functioning normally before conducting the test and then verify that the voltmeter is still functioning normally after conducting the test. The inspection should verify the following information:

- The protective fuse is good.

- The case/enclosure is free from cracks and is not otherwise broken.

- The readout is clear and legible.

- The insulation on the leads is complete and undamaged.

- The shroud on the banana plug is complete and undamaged.

- The finger guards are in place.

- The retractable probe covers are in place and functional.

- Each lead is continuous.

⚡ ENERGY CONTROL PROCEDURES

NFPA 70E defines three techniques for controlling hazardous energy:

1. Individual qualified employee control

2. Simple lockout–tagout

3. Complex lockout–tagout

For more detailed information, see the following:

• *NFPA 70E: Handbook for Electrical Safety in the Workplace*

• *NECA Guide to NFPA 70E Lockout–Tagout Requirements*

Individual Qualified Employee Control

This method is permitted when electrical equipment with exposed conductors and circuit parts is deenergized for minor maintenance (e.g., adjusting, servicing, cleaning, or inspection). The individual qualified employee control method of controlling energy can be performed without placing lockout–tagout devices when *all* of the following requirements are met:

• The disconnecting means is near the equipment and continuously visible to the qualified employee.

• The disconnecting means is continuously within arm's reach.

• The work does not extend beyond one shift.

Simple Lockout–Tagout Procedure

This method is permitted when a single disconnect serves as the energy isolating device (e.g., when the circuit breaker feeding a branch circuit is not within sight of the equipment being worked on) **(see Figure 7)**:

FIGURE 7   Simple lockout devices. (Photo courtesy of Grainger)

- A written lockout–tagout plan is not required.

- Each worker is responsible for his or her own lockout–tagout.

Complex Lockout–Tagout Procedure

Complex lockout–tagout is required under one or more of the following circumstances:

- Multiple energy sources and/or disconnects are involved.

- Multiple crews, crafts, or employers are involved.

- Work is going on simultaneously at different locations.

- Particular sequences of deenergizing are needed.

- A job or task continues for more than one shift.

 ENERGY CONTROL PROCEDURES

Power sources controlled by a complex lockout–tagout procedure normally have more than one padlock and/or tag **(see Figure 8)**.

FIGURE 8 Complex lockout.

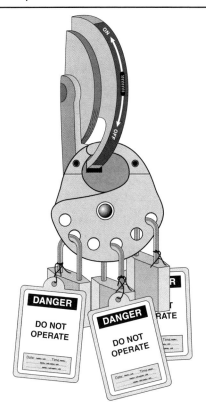

■ SAFETY PRACTICES WHEN ONE CONDUCTOR IS OR MIGHT BE ENERGIZED

Safe work practices for working within the limited approach boundary of energized or potentially energized electrical conductors or circuit parts consist of the following:

- Energized Electrical Work Permit

- Shock approach boundaries

- Arc-flash protection boundary

- PPE

- FR clothing

⚡ SAFE ELECTRICAL WORK PRACTICES

Sometimes it is not feasible to establish an *electrically safe work condition* (turn off the power) before doing repair or maintenance work on electrical systems. For instance, taking a load reading with a clamp-on ammeter requires the circuit to be energized. When working on exposed conductors and equipment operating at 50 to 600 volts, *NFPA 70E* requires the following:

• A hazard/risk analysis

• An authorized Energized Electrical Work Permit

• Determination of the shock approach boundaries hazard/risk analysis (shock hazard analysis)

• Determination of the arc-flash protection boundary hazard/risk analysis (arc-flash hazard analysis). Qualified persons must use appropriate PPE when working inside the arc-flash protection boundary.

• Determination of the hazard/risk category (HRC) or the incident energy for the task to be performed. Different types of PPE, including FR clothing, are needed for different hazard/risk categories.

Other safety precautions include the following:

• Workers must not reach blindly into areas that might contain exposed energized electrical conductors or circuit parts.

• Workers must ensure that they have adequate illumination for energized work.

• Workers must not perform energized work when their view is obstructed.

• Workers must not wear conductive items such as jewelry, watchbands, unrestrained metal frame glasses, clothing with metallic threads, or metallic body piercings. Steel-toed work boots required by OSHA regulations are acceptable because leather or fabric covers the metal.

![icon] SAFE ELECTRICAL WORK PRACTICES

Energized Electrical Work Permit

The first electrical safe work practice when working on energized electrical conductors or circuit parts is to *justify* the reason that work must be performed with the power on. The person authorizing the work (management, safety officer, or owner) must sign an Energized Electrical Work Permit like the one shown in *NFPA 70E-2009*, Annex J, Figure J.1. This permit contains the following:

- Description of the work to be done

- Justification of why the work must be performed with the circuit/equipment energized (see the discussion following this list)

- Summary of the hazard/risk analysis used to determine approach boundaries (shock protection) and the arc-flash protection boundary

- List of the PPE to be used, including FR clothing

- Description of methods to be used to keep unqualified persons out of the work area

- Requirement for a job briefing before the work is performed

- The signature(s) of the person(s) authorizing the work to be performed with the circuit/equipment energized

NOTE: Energized Electrical Work Permits often are called Live Work Permits in the field.

Reasons that Justify Working with the Circuit/Equipment Energized

Additional or Increased Hazards

NFPA 70E allows energized work to be performed if deenergizing introduces additional or increased hazards. Examples of such hazards include interrupting life support equipment, deactivating emergency alarm systems, and shutting down ventilation equipment in hazardous (classified) locations.

◨ SAFE ELECTRICAL WORK PRACTICES

Design or Operational Limitations

NFPA 70E allows energized work to be performed if deenergizing is infeasible due to equipment design or operational limitations. Examples of such infeasibility include testing and troubleshooting work that can be done only while the circuit/equipment is energized, or process manufacturing such as steel and glass making where power cannot be shut down without damaging the equipment.

The Difference Between *INCONVENIENT* and *INFEASIBLE*

The Task

Ballasts in fluorescent ceiling fixtures in an office need to be replaced. The customer does not want to turn off the branch circuit(s) feeding the fixtures during normal working hours for fear of inconveniencing workers and reducing productivity.

The Best Solution

The best solution is to replace the ballasts at night or over a weekend, when an *electrically safe work condition* can be established and electricians can work using only minimal PPE (eye protection and work gloves).

REMEMBER: Turning off the power is always the safest way to work!

Acceptable Alternate Solution

Electrical workers can have the customer complete an Energized Electrical Work Permit. Note that the work must be justified as defined by *NFPA 70E*. The workers can then do the work while the circuit is energized, during normal office hours, with the electricians using PPE and following all other safety precautions required by *NFPA 70E*.

⏻ SAFE ELECTRICAL WORK PRACTICES

This is an *acceptable* solution because it complies with *NFPA 70E* and provides a degree of protection for workers. However, turning off the power is always the safest way to work on electrical conductors and equipment.

NOTE: Working energized is a last resort, and you always need PPE then.

Determine Shock Approach Boundaries

The second electrical safe work practice when working with the circuit/equipment energized is determining the shock approach boundaries. These boundaries help protect against shock and electrocution. Approach boundaries are identified as limited, restricted, and prohibited **(see Figure 9)**. Crossing one of these approach boundaries increases the chance that a worker might contact an exposed energized electrical conductor or circuit part.

FIGURE 9   Approach boundaries.

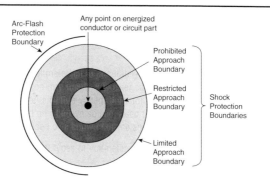

Arc-Flash Protection Boundary

Any point on energized conductor or circuit part

Prohibited Approach Boundary

Restricted Approach Boundary

Shock Protection Boundaries

Limited Approach Boundary

⬛ SAFE ELECTRICAL WORK PRACTICES

For energized electrical systems operating at 50 to 600 volts, approach boundaries are as follows:

- *Limited Approach Boundary*—This boundary is larger for movable conductors than for fixed circuit parts.

- *Restricted Approach Boundary*—This boundary includes an *inadvertent movement adder*. It allows for the fact that a person's hand or tool might slip, or someone else might jostle the worker from behind.

- *Prohibited Approach Boundary*—This boundary is considered the same as touching an exposed energized circuit part. **(See Table 2.)**

Limited Approach Boundary

- *Qualified Persons.* Only electrically qualified persons can work inside a limited approach boundary without special precautions.

- *Unqualified Persons.* Persons such as management employees, cleaning crews, painters, and other construction trades can enter the limited approach boundary only when escorted and advised by an electrically qualified person. Otherwise, unqualified persons must be kept outside a limited approach boundary by barricades and warning signs.

TABLE 2   Approach Boundaries

Phase-to-Phase Voltage	Limited Approach Boundary Movable	Fixed	Restricted Approach Boundary	Prohibited Approach Boundary
50 to 300	10 feet	42 inches	Avoid contact	Avoid contact
301 to 600	10 feet	42 inches	1 foot	1 inch

⬛ SAFE ELECTRICAL WORK PRACTICES

Restricted Approach Boundary

- *Qualified Persons.* Electrically qualified persons can enter a restricted approach boundary only when wearing and using PPE, such as adequately rated gloves and tools, for protection from shock.

- *Unqualified Persons.* These persons are not permitted within a restricted approach boundary.

Prohibited Approach Boundary

- *Qualified Persons* can enter a prohibited approach boundary only when using the same precautions as if they were in direct contact with an exposed energized electrical conductor or circuit part. Voltage testing is considered a task that requires entering the prohibited approach boundary.

- *Unqualified Persons* are not permitted within a prohibited approach boundary.

Not Flash Protection!

Limited, restricted, and prohibited approach boundaries are for shock protection only. They have nothing to do with protecting workers from arc-flash and arc-blast hazards, or selecting FR clothing. See the following section.

Arc-Flash Protection Boundary (Arc-Flash and Arc-Blast Protection)

- The default arc-flash protection boundary is 4 feet for energized electrical systems operating at 600 volts or less, when the product of the clearing time and available fault current does not exceed 100 kA-cycles.

- Qualified persons working inside the arc-flash protection boundary must wear appropriate PPE, including FR clothing.

⬛ SAFE ELECTRICAL WORK PRACTICES

Determining Available Fault Current

The level of fault current available in an electrical distribution system often can be furnished by the electric utility supplying the building or structure, or by the facility engineer/manager. Levels of available fault current also can be determined as follows:

• Calculate the AIC by hand, as shown in Annex D of *NFPA 70E*.

• Use arc-flash calculation software.

• Hire an engineering/safety consultant.

NOTE: These methods are outside the scope of this guide. For electrical systems with more than 50,000 amperes of fault current available, or where this information cannot be furnished by the local utility or facility engineer, see the NFPA 70E *standard.*

Turning off the power is always the safest way to do electrical construction and maintenance work. Workers still need the correct PPE to do the "test before touch" part of getting to an electrically safe condition. Once an electrically safe work condition has been established, it is not necessary to determine the available fault current, because an arc-flash protection boundary does not exist. Qualified persons can work on deenergized systems using minimal PPE, such as hard hats, work gloves, and safety goggles or glasses.

⚡ SAFE ELECTRICAL WORK PRACTICES

Determine the Hazard/Risk Category for the Task

NFPA 70E defines five hazard/risk categories (HRC) based on the available incident energy when the circuit/equipment energized. The HRC is used to determine the following:

1. Whether or not voltage-rated gloves must be used

2. Whether or not voltage-rated tools must be used

3. The type of FR clothing that must be worn

NFPA 70E standard defines five hazard/risk categories for work within the arc-flash protection boundary.

Systems Rated at 600 Volts and Less

All work on most electrical systems operating at 600 volts and less falls into HRC 3 and below. Generally, HRC 4 applies only to systems operating at 1000 volts and more. This *Ugly's* guide includes information on Hazard/Risk Categories 1, 2, 3, and 4.

How to Determine the HRC Category for Each Work Task

• **See Table 3** in this *Ugly's* guide for the HRC for common electrical tasks.

• Always confirm that the power system parameters meet the limits set in the table notes.

• The listing for each task also states whether or not voltage-rated gloves must be worn and whether or not voltage-rated tools must be used.

⏻ SAFE ELECTRICAL WORK PRACTICES

REMEMBER: Hazard/risk categories apply only when the equipment or circuit is energized. If an electrically safe work condition has been established (i.e., if the power has been turned off and confirmed off by test using correct PPE when performing the test), there is no HRC.

Select the Appropriate PPE, Including FR Clothing

- Once the worker has selected the appropriate HRC for a task from *NFPA 70E* Table {C-9}, he or she must use *NFPA 70E* Table {C-10} to select the appropriate PPE and FR clothing.

- Next, the worker must use *NFPA 70E* Table {C-11} to select a complete FR clothing system that consists of several layers of clothing.

ADVANTAGES OF WORKING DEENERGIZED

TABLE 3   Table 130.7(c)(9)   Hazard/Risk Category Classifications and Use of Rubber Insulating Gloves and Insulated and Insulating Hand Tools

Tasks Performed on Energized Equipment	Hazard/Risk Category	Rubber Insulating Gloves	Insulated and Insulating Hand Tools
Panelboards or Other Equipment Rated 240 V and Below — Note 1			
Perform infrared thermography and other non-contact inspections outside the restricted approach boundary	0	N	N
Circuit breaker (CB) or fused switch operation with covers on	0	N	N
CB or fused switch operation with covers off	0	N	N
Work on energized electrical conductors and circuit parts, including voltage testing	1	Y	Y
Remove/install CBs or fused switches	1	Y	Y
Removal of bolted covers (to expose bare, energized electrical conductors and circuit parts)	1	N	N
Opening hinged covers (to expose bare, energized electrical conductors and circuit parts)	0	N	N
Work on energized electrical conductors and circuit parts of utilization equipment fed directly by a branch circuit of the panelboard	1	Y	Y

# ADVANTAGES OF WORKING DEENERGIZED

TABLE 3 Hazard/Risk Category Classifications—continued

Tasks Performed on Energized Equipment	Hazard/Risk Category	Rubber Insulating Gloves	Insulated and Insulating Hand Tools
Panelboards or Switchboards Rated >240 V and up to 600 V (with molded case or insulated case circuit breakers) — Note 1			
Perform infrared thermography and other non-contact inspections outside the restricted approach boundary	1	N	N
CB or fused switch operation with covers on	0	N	N
CB or fused switch operation with covers off	1	Y	N
Work on energized electrical conductors and circuit parts, including voltage testing	2*	Y	Y
Work on energized electrical conductors and circuit parts of utilization equipment fed directly by a branch circuit of the panelboard or switchboard	2*	Y	Y
600 V Class Motor Control Centers (MCCs) — Note 2 (except as indicated)			
Perform infrared thermography and other non-contact inspections outside the restricted approach boundary	1	N	N
CB or fused switch or starter operation with enclosure doors closed	0	N	N
Reading a panel meter while operating a meter switch	0	N	N

(*continues*)

ADVANTAGES OF WORKING DEENERGIZED

TABLE 3   Hazard/Risk Category Classifications—continued

Tasks Performed on Energized Equipment	Hazard/Risk Category	Rubber Insulating Gloves	Insulated and Insulating Hand Tools
CB or fused switch or starter operation with enclosure doors open	1	N	N
Work on energized electrical conductors and circuit parts, including voltage testing	2*	Y	Y
Work on control circuits with energized electrical conductors and circuit parts 120 V or below, exposed	0	Y	Y
Work on control circuits with energized electrical conductors and circuit parts >120 V, exposed	2*	Y	Y
Insertion or removal of individual starter "buckets" from MCC — Note 3	4	Y	N
Application of safety grounds, after voltage test	2*	Y	N
Removal of bolted covers (to expose bare, energized electrical conductors and circuit parts) — Note 3	4	N	N
Opening hinged covers (to expose bare, energized electrical conductors and circuit parts) — Note 3	1	N	N
Work on energized electrical conductors and circuit parts of utilization equipment fed directly by a branch circuit of the motor control center	2*	Y	Y

ADVANTAGES OF WORKING DEENERGIZED

TABLE 3   Hazard/Risk Category Classifications—continued

Tasks Performed on Energized Equipment	Hazard/Risk Category	Rubber Insulating Gloves	Insulated and Insulating Hand Tools
600 V Class Switchgear (with power circuit breakers or fused switches) — Note 4			
Perform infrared thermography and other non-contact inspections outside the restricted approach boundary	2	N	N
CB or fused switch operation with enclosure doors closed	0	N	N
Reading a panel meter while operating a meter switch	0	N	N
CB or fused switch operation with enclosure doors open	1	N	N
Work on energized electrical conductors and circuit parts, including voltage testing	2*	Y	Y
Work on control circuits with energized electrical conductors and circuit parts 120 V or below, exposed	0	Y	Y
Work on control circuits with energized electrical conductors and circuit parts >120 V, exposed	2*	Y	Y
Insertion or removal (racking) of CBs from cubicles, doors open or closed	4	N	N
Application of safety grounds, after voltage test	2*	Y	N
Removal of bolted covers (to expose bare, energized electrical conductors and circuit parts)	4	N	N

(*continues*)

43

 **ADVANTAGES OF WORKING DEENERGIZED**

TABLE 3   Hazard/Risk Category Classifications—continued

Tasks Performed on Energized Equipment	Hazard/Risk Category	Rubber Insulating Gloves	Insulated and Insulating Hand Tools
Opening hinged covers (to expose bare, energized electrical conductors and circuit parts)	2	N	N
Other 600 V Class (277 V through 600 V, nominal) Equipment — Note 2 (except as indicated)			
Lighting or small power transformers (600 V, maximum)			
Removal of bolted covers (to expose bare, energized electrical conductors and circuit parts)	2*	N	N
Opening hinged covers (to expose bare, energized electrical conductors and circuit parts)	1	N	N
Work on energized electrical conductors and circuit parts, including voltage testing	2*	Y	Y
Application of safety grounds, after voltage test	2*	Y	N
Revenue meters (kW-hour, at primary voltage and current) Insertion or removal	2*	Y	N
Cable trough or tray cover removal or installation	1	N	N
Miscellaneous equipment cover removal or installation	1	N	N
Work on energized electrical conductors and circuit parts, including voltage testing	2*	Y	Y

ADVANTAGES OF WORKING DEENERGIZED

TABLE 3 Hazard/Risk Category Classifications—continued

Tasks Performed on Energized Equipment	Hazard/Risk Category	Rubber Insulating Gloves	Insulated and Insulating Hand Tools
Application of safety grounds, after voltage test	2*	Y	N
Insertion or removal of plug-in devices into or from busways	2*	Y	N
NEMA E2 (fused contactor) Motor Starters, 2.3 kV Through 7.2 kV			
Perform infrared thermography and other non-contact inspections outside the restricted approach boundary	3	N	N
Contactor operation with enclosure doors closed	0	N	N
Reading a panel meter while operating a meter switch	0	N	N
Contactor operation with enclosure doors open	2*	N	N
Work on energized electrical conductors and circuit parts, including voltage testing	4	Y	Y
Work on control circuits with energized electrical conductors and circuit parts 120 V or below, exposed	0	Y	Y
Work on control circuits with energized electrical conductors and circuit parts >120 V, exposed	3	Y	Y
Insertion or removal (racking) of starters from cubicles, doors open or closed	4	N	N

(continues)

ADVANTAGES OF WORKING DEENERGIZED

TABLE 3   Hazard/Risk Category Classifications—continued

Tasks Performed on Energized Equipment	Hazard/Risk Category	Rubber Insulating Gloves	Insulated and Insulating Hand Tools
Application of safety grounds, after voltage test	3	Y	N
Removal of bolted covers (to expose bare, energized electrical conductors and circuit parts)	4	N	N
Opening hinged covers (to expose bare, energized electrical conductors and circuit parts)	3	N	N
Insertion or removal (racking) of starters from cubicles of arc-resistant construction, tested in accordance with IEEE C37.20.7, doors closed only	0	N	N
Metal Clad Switchgear, 1 kV Through 38 kV			
Perform infrared thermography and other non-contact inspections outside the restricted approach boundary	3	N	N
CB operation with enclosure doors closed	2	N	N
Reading a panel meter while operating a meter switch	0	N	N
CB operation with enclosure doors open	4	N	N
Work on energized electrical conductors and circuit parts, including voltage testing	4	Y	Y
Work on control circuits with energized electrical conductors and circuit parts 120 V or below, exposed	2	Y	Y

46

ADVANTAGES OF WORKING DEENERGIZED

TABLE 3   Hazard/Risk Category Classifications—continued

Tasks Performed on Energized Equipment	Hazard/Risk Category	Rubber Insulating Gloves	Insulated and Insulating Hand Tools
Work on control circuits with energized electrical conductors and circuit parts >120 V, exposed	4	Y	Y
Insertion or removal (racking) of CBs from cubicles, doors open or closed	4	N	N
Application of safety grounds, after voltage test	4	Y	N
Removal of bolted covers (to expose bare, energized electrical conductors and circuit parts)	4	N	N
Opening hinged covers (to expose bare, energized electrical conductors and circuit parts)	3	N	N
Opening voltage transformer or control power transformer compartments	4	N	N
Arc-Resistant Switchgear Type 1 or 2 (for clearing times of <0.5 sec with a perspective fault current not to exceed the arc resistant rating of the equipment)			
CB operation with enclosure door closed	0	N	N
Insertion or removal (racking) of CBs from cubicles, doors closed	0	N	N
Insertion or removal of CBs from cubicles with door open	4	N	N

(*continues*)

ADVANTAGES OF WORKING DEENERGIZED

TABLE 3   Hazard/Risk Category Classifications—continued

Tasks Performed on Energized Equipment	Hazard/Risk Category	Rubber Insulating Gloves	Insulated and Insulating Hand Tools
Work on control circuits with energized electrical conductors and circuit parts 120 V or below, exposed	2	Y	Y
Insertion or removal (racking) of ground and test device with door closed	0	N	N
Insertion or removal (racking) of voltage transformers on or off the bus door closed	0	N	N
Other Equipment 1 kV Through 38 kV			
Metal-enclosed interrupter switchgear, fused or unfused			
Switch operation of arc-resistant-type construction, tested in accordance with IEEE C37.20.7, doors closed only	0	N	N
Switch operation, doors closed	2	N	N
Work on energized electrical conductors and circuit parts, including voltage testing	4	Y	Y
Removal of bolted covers (to expose bare, energized electrical conductors and circuit parts)	4	N	N
Opening hinged covers (to expose bare, energized electrical conductors and circuit parts)	3	N	N
Outdoor disconnect switch operation (hookstick operated)	3	Y	Y

TABLE 3   Hazard/Risk Category Classifications—continued

Tasks Performed on Energized Equipment	Hazard/Risk Category	Rubber Insulating Gloves	Insulated and Insulating Hand Tools
Outdoor disconnect switch operation (gang-operated, from grade)	2	Y	N
Insulated cable examination, in manhole or other confined space	4	Y	N
Insulated cable examination, in open area	2	Y	N

General Notes (applicable to the entire table):
(a) Rubber insulating gloves are gloves rated for the maximum line-to-line voltage upon which work will be done.
(b) Insulated and insulating hand tools are tools rated and tested for the maximum line-to-line voltage upon which work will be done, and are manufactured and tested in accordance with ASTM F 1505, Standard Specification for Insulated and Insulating Hand Tools.
(c) Y = yes (required), N = no (not required).
(d) For systems rated less than 1000 volts, the fault currents and upstream protective device clearing times are based on an 18 in. working distance.
(e) For systems rated 1 kV and greater, the Hazard/Risk Categories are based on a 36 in. working distance.
(f) For equipment protected by upstream current limiting fuses with arcing fault current in their current limiting range (½ cycle fault clearing time or less), the hazard/risk category required may be reduced by one number.

(notes continued)

ADVANTAGES OF WORKING DEENERGIZED

Specific Notes (as referenced in the table):

1. *Maximum of 25 kA short circuit current available; maximum of 0.03 sec (2 cycle) fault clearing time.*
2. *Maximum of 65 kA short circuit current available; maximum of 0.03 sec (2 cycle) fault clearing time.*
3. *Maximum of 42 kA short circuit current available; maximum of 0.33 sec (20 cycle) fault clearing time.*
4. *Maximum of 35 kA short circuit current available; maximum of up to 0.5 sec (30 cycle) fault clearing time.*

⚡ TYPES OF PROTECTION

Gloves

- Gloves increase the resistance of the current path between an energized electrical conductor and a person's skin.

- By increasing the impedance, the gloves reduce the amount of current that flows through the person's body as a result of a contact (accidental or intentional).

- The amount of current that flows through body tissue is directly proportional to the kind and degree of injury.

If the insulation value of a worker's glove is overcome by the voltage potential of the electrical conductor, the glove resistance decreases to the point that dangerous currents might flow through the worker's body. Workers must have the ability to select an insulating glove that cannot be affected by the voltage potential of the circuit.

Hazards

- The rubber used in the insulating layer can burn, but it is difficult to ignite.

- The leather protectors offer significant protection from the thermal hazard associated with an arcing fault.

- Although leather protectors have no specific incident energy rating, experience shows that they offer significant protection from exposure to an arcing fault **(see Figure 10)**.

Size and Style

Voltage-rated gloves should fit the hands of the person wearing them. Glove size for a specific hand is determined by measuring the circumference of the palm at its widest point. The palm measurement is the size that should be ordered **(see Figure 11)**.

TYPES OF PROTECTION

FIGURE 10 **Gloves.** (Courtesy of Salisbury Electrical Safety LLC)

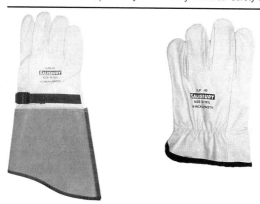

FIGURE 11 Measuring the hand for glove fit.

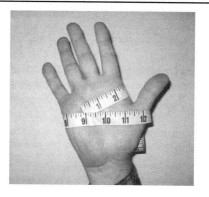

TYPES OF PROTECTION

Leather Protectors

- The function of leather protectors, sometimes called leathers, is to reduce the chance of damage to the insulating glove.

- When a worker touches an energized conductor with the gloves, the entire leather protector is energized at the same voltage of the conductor.

- Therefore, the leather protector must not contact any uninsulated body part.

- The insulating rubber glove must extend beyond the leather protector by a length that is sufficient to eliminate the chance of creepage between the leather protector and the worker's skin.

- Gloves insulate the hands of the worker from current flow (shock) because the rubber is an insulator.

- Leather is not an insulator; therefore, leather protectors protect only the gloves, not the hands. In an arc flash, however, leather protectors become the protecting component **(see Figure 12)**.

FIGURE 12   Leather protectors. (Courtesy of Salisbury Electrical Safety LLC)

 TYPES OF PROTECTION

Selection and Use of Voltage-Rated Gloves

- Rubber gloves used for shock protection should be called rated gloves or voltage-rated (insulating) gloves.

- Although many different types of rubber gloves are on the market, unless they meet consensus requirements as voltage-rated (insulating) gloves, they should never be used for shock protection (**see Figure 13**).

FIGURE 13   Voltage-rated gloves. (Courtesy of Salisbury Electrical Safety LLC)

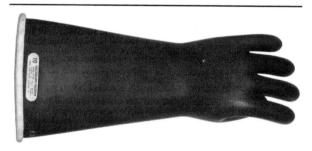

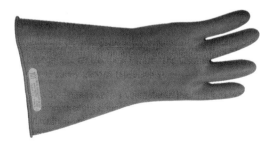

⌂ TYPES OF PROTECTION

Inspection

- Gloves and leather protectors must be inspected before each use **(see Figure 14 and Figure 15)**. For a flow chart for glove inspection, **see Figure 16**.

- Although nationally recognized standards require visual examination and inspection of both the insulating glove and the leather protector

FIGURE 14 Damaged gloves. (Courtesy of Salisbury Electrical Safety LLC)

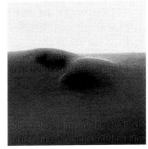

TYPES OF PROTECTION

FIGURE 15 **Glove inspection.** (Courtesy of Salisbury Electrical Safety LLC)

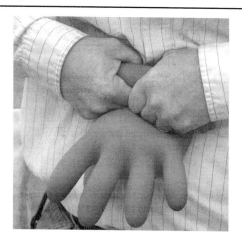

before use, the real reason for inspecting them is to prevent electrocution.

- Any damage to either the insulating material or the leather protectors can result in the wearer being electrocuted when direct contact is made with an exposed energized conductor.

Sleeves

- Sleeves serve to increase the impedance of the current path between an exposed energized conductor and a person attempting to manipulate the conductor or another circuit part **(see Figure 17)**.

TYPES OF PROTECTION

FIGURE 16   Glove inspection flow chart.

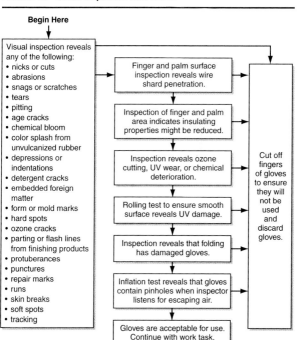

Begin Here

Visual inspection reveals any of the following:
- nicks or cuts
- abrasions
- snags or scratches
- tears
- pitting
- age cracks
- chemical bloom
- color splash from unvulcanized rubber
- depressions or indentations
- detergent cracks
- embedded foreign matter
- form or mold marks
- hard spots
- ozone cracks
- parting or flash lines from finishing products
- protuberances
- punctures
- repair marks
- runs
- skin breaks
- soft spots
- tracking

Finger and palm surface inspection reveals wire shard penetration.

Inspection of finger and palm area indicates insulating properties might be reduced.

Inspection reveals ozone cutting, UV wear, or chemical deterioration.

Rolling test to ensure smooth surface reveals UV damage.

Inspection reveals that folding has damaged gloves.

Inflation test reveals that gloves contain pinholes when inspector listens for escaping air.

Gloves are acceptable for use. Continue with work task.

Cut off fingers of gloves to ensure they will not be used and discard gloves.

 **TYPES OF PROTECTION**

FIGURE 17 **Worker wearing sleeves correctly.** (© David Frazier Photolibrary, Inc./Alamy Images)

- By increasing the impedance, the sleeves reduce the amount of current flow to a predetermined level.

- The objective is to reduce the current flow to a level that cannot harm the person wearing the sleeves. Sleeves must never be worn without voltage-rated gloves.

- Sleeves are held in place by buttons and straps or a harness **(see Figure 18)**.

TYPES OF PROTECTION

FIGURE 18 Straps and harness used to hold sleeves in place.
(Courtesy of Salisbury Electrical Safety LLC)

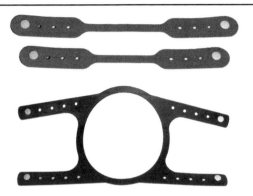

Mats and Matting

- A rubber mat is essentially a subset of matting **(see Figure 19)**.

- Mats are intended to be installed on the floor in front of electrical equipment.

FIGURE 19 Matting. (Courtesy of Salisbury Electrical Safety LLC)

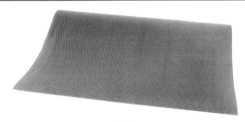

⚡ TYPES OF PROTECTION

How Mats and Matting Provide Protection

- Muscle tissue reacts to current flow through the body. Shock and electrocution are the result of current flow.

- Voltage acts as the pressure that forces the current to flow. As the amount of current increases, the body reacts more violently.

- If the current flow can be limited to a value that has no deleterious effect, then neither shock nor electrocution are possible.

- The intended role of mats and matting is to reduce the amount of current flow should contact with an exposed energized conductor be made.

If a worker touches an exposed energized conductor (and no other conductor or grounded surface) while standing on an appropriately rated mat, the amount of current flow through the worker's body will be small, and the worker will not be injured.

Rubber gloves and blankets insert insulation between a worker and an energized component. The worker can be in contact with earth ground and still current cannot flow, because the gloves or blanket reduce the chance of contact with an energized component **(see Figure 20)**.

Blankets

In normal use, blankets come into direct contact with the electrical conductor. Consequently, the prohibited approach boundary is penetrated each time a blanket is installed.

- Current consensus standards require each employer to determine necessary procedural and administrative actions when the prohibited approach boundary is penetrated.

- The circuit should be deenergized before the blanket(s) is (are) installed.

TYPES OF PROTECTION

FIGURE 20 Rubber gloves and mats add resistance between worker and earth ground.

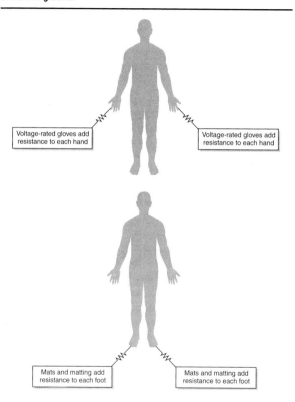

Voltage-rated gloves add resistance to each hand

Voltage-rated gloves add resistance to each hand

Mats and matting add resistance to each foot

Mats and matting add resistance to each foot

⚡ TYPES OF PROTECTION

- The worker should ensure that the circuit is clear of any potential fault prior to reenergizing the circuit.

It is sometimes not possible to deenergize the circuit to install blankets for an emergency work task. In that instance, strict adherence to procedural requirements is critical.

- Blankets normally are wrapped around an electrical conductor in such a way that the worker is less likely to contact the conductor when executing the work task **(see Figure 21)**.

- The blankets may be held in place by non-conductive cable ties, buttons, or clamp pins. Electricians frequently refer to the clamp pins as clothespins, because they have the general appearance of large clothespins. Blankets with hook-and-pile (Velcro®) fastening along the edges are also available.

- Blankets are intended to provide temporary insulation on an electrical conductor and should not be left in place after the work task is complete.

FIGURE 21   Blanket. (Courtesy of Salisbury Electrical Safety LLC)

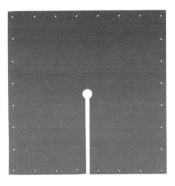

⚡ TYPES OF PROTECTION

Hazards

Insulating blankets are ineffective as protection from other electrical hazards such as arc flash or arc blast. Rubber blankets are held in place by clothespins, cable ties, or other means, and the device used to hold the blanket in place could become a missile should an arcing fault occur, thus introducing a new hazard.

An arcing fault is electrical current flowing in air between two conductors.

- Arcing faults may be the result of component failure.

- However, a worker making inadvertent contact with an energized conductor usually initiates an arcing fault.

- An adequately rated blanket reduces the chance of initiating a fault on the conductor. However, the blanket will have no effect in reducing a worker's exposure to the thermal and blast effects of an arcing fault.

An arcing fault also produces a significant pressure wave. Fasteners could become projectiles should an arcing fault occur. Placement of fasteners is an important consideration when positioning the fasteners on the temporary blanket. Buttons are more likely to cause injury than cable ties **(see Figure 22 and Figure 23)**.

FIGURE 22   Buttons hold rubber blankets in place. (Courtesy of Salisbury Electrical Safety LLC)

TYPES OF PROTECTION

FIGURE 23 Clothespins hold rubber blankets in place. (Courtesy of Salisbury Electrical Safety LLC)

Footwear

For current to flow through a worker's body, the worker must contact an exposed energized conductor and another conductor, such as earth ground. If the worker wears footwear that is adequately rated, foot contact with earth ground is reduced or eliminated.

Dielectric footwear must be constructed as defined in the standard *ASTM F2413* **(see Figure 24)**. Tests to determine the electrical protective nature of dielectric footwear are defined in *ASTM F1116, Standard Test Method for Determining Dielectric Strength of Dielectric Footwear.* These national consensus standards provide assurance that the footwear meets normal expectations for protecting a worker's feet from both physical and electrical shock hazard.

Live-Line Tools (Hot Sticks)

- Electricians have used the term "hot stick" for many years. The term suggests that the device is intended to contact an energized conductor.

- When an electrician says that a conductor is "hot," he or she probably is talking about the fact that a conductor is energized relative to earth ground. However, the electrician could be talking about the thermal condition of a conductor that is overloaded.

TYPES OF PROTECTION

FIGURE 24 Dielectric footwear. (Courtesy of Salisbury Electrical Safety LLC)

- In another example, when thermographers perform analyses on electrical conductors and terminations, the term "hot" refers to the thermal condition of the conductor or termination.

Use of the term "hot" to mean both "energized" and "high thermal temperature" can be very confusing. The term "hot stick," then, is a misnomer, although it is the most commonly used term.

- OSHA regulations use the term "hot" to refer to a thermal condition instead of an energized condition.

- In the OSHA regulations, the term "hot stick" is used only in an appendix, and then only in an attempt to ensure complete understanding by the user. In *29 CFR 1910.269(j)*, OSHA uses the descriptive term "live-line" tools instead of "hot stick" to discuss tools constructed from fiberglass-reinforced plastic (FRP).

 **TYPES OF PROTECTION**

- In common usage, both "hot stick" and "live-line tools" refer to a family of tools constructed from FRP material that are used to perform specific actions on an energized conductor or component **(see Figure 25 and Figure 26)**.

Hazards

- Should an arcing fault occur when the worker is using a live-line tool, the worker could be exposed to the thermal hazard associated with the arcing fault.

- The live-line tool provides no protection from this hazard.

FIGURE 25 **"Shot gun" live line tool.** (Courtesy of Salisbury Electrical Safety LLC)

FIGURE 26 **Live-line tool construction.**

 **TYPES OF PROTECTION**

- The chance of initiating an arcing fault by contacting an energized component with a live-line tool is remote. However, equipment or devices that may be attached to the end of a live-line tool can initiate an arcing fault **(see Figure 27)**.

- The length of the live-line tool is the critical component of preventing thermal injury to an otherwise unprotected worker.

FIGURE 27   Arcing fault on an overhead line. (Courtesy of D. Ray Crow)

⊔ TYPES OF PROTECTION

Selection and Use

- Customers expect that utilities will keep electrical energy available for their use in homes, businesses, commercial endeavors, and industrial locations.

- Utilities and manufacturers respond to that demand by developing equipment, tools, and work processes that enable maximum system availability for customers. Customers also expect that utilities will provide the electricity at minimum costs.

- Utilities respond by developing equipment that minimizes the cost of the installation to the consumer while maintaining the integrity of the service. For instance, fuse clips become disconnecting devices. Devices that permit an appropriately assembled live-line tool to operate the disconnecting device replace ground-operated switching devices.

- Frequently, the means to disconnect the utility will be located at the top of a pole **(see Figure 28)**. A live-line tool enables a worker to operate the disconnecting means from the ground or from a bucket truck.

FIGURE 28   Live-line tool disconnect switch.

⚡ TYPES OF PROTECTION

Disconnecting means that are operated by a live-line tool cost much less than gang-operated switches **(see Figure 29)**. Gang-operated means that all phases are connected together mechanically so that all phases break and make together.

- Generally, larger manufacturing facilities require immediate access to disconnecting means. These facilities normally install gang-operated switches, enclosed fusible switches, or circuit breakers in an enclosure.

- This installation enables workers to operate the handle of the equipment quickly to deenergize the electrical service. However, some industrial facilities employ pad-mounted transformers with disconnecting devices that are operated using live-line tools **(see Figure 30)**.

- Thus, industrial workers must follow the same work processes and practices as utility workers.

FIGURE 29 Operating a disconnect switch. (© Matthew Collingwood/ShutterStock, Inc.)

TYPES OF PROTECTION

FIGURE 30   Pad-mounted transformer.

Accessories are available that can equip the live-line tool to perform most functions. For example, the live-line tool can be utilized as a tool for rescue workers who might be in contact with an energized conductor, or the tool can be equipped with grounding equipment to discharge static voltage **(see Figure 31)**.

FR Clothing

Members of the *NFPA 70E* Technical Committee agreed that eliminating the risk of clothing ignition or melting would prevent many injuries and reduce the severity of the remainder. The committee also agreed that the flame-resistant clothing available at the time would reduce the severity of burn injuries. At the very least, wearing (arc-rated) FR clothing would eliminate the risk associated with clothing ignition.

TYPES OF PROTECTION

FIGURE 31 Static discharge stick. (Courtesy of Salisbury Electrical Safety LLC)

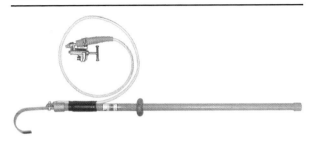

Clothing that meets the requirements of the current edition of *ASTM F1506, Standard Performance Specification for Flame-Resistant Textile Materials for Wearing Apparel for Use by Electrical Workers Exposed to Momentary Electric Arc and Related Thermal Hazards* is considered FR clothing. Fabric or other material that is flame resistant does not ignite. The flame-resistant characteristic remains with the clothing for the life of the garment. For a list of standards that regulate arc-flash protective equipment, **see Table 4**.

Consensus requirements for FR clothing do not limit the materials of construction to particular fabrics or materials.

- In addition to materials that are inherently flame resistant, cotton and other fabrics can be treated with an agent to temporarily change the ignition characteristic of the fabric.

- Generally, the agent is a chemical that imparts a characteristic that retards the spread of flame after ignition.

 **TYPES OF PROTECTION**

TABLE 4   Standards for Arc-Flash Protection

Subject	Number and Title
Apparel	• ASTM D6413, Standard Test Method for Flame-Resistance of Textiles • ASTM F1449, Standard Guide for Care and Maintenance of Flame-Resistant Clothing • ASTM F1506, Standard Performance Specification for Flame-Resistant Textile Materials for Wearing Apparel for Use by Electrical Workers Exposed to Momentary Electric Arc and Related Thermal Hazards • ASTM F1958, Standard Test Method for Determining the Ignitability of Non-Flame-Resistance Materials for Clothing by Electric Arc Exposure Method Using Mannequins • ASTM F1959, Standard Test Method for Determining the Arc Thermal Performance Value of Materials for Clothing
Gloves and sleeves	• ASTM D120, Standard Specification for Rubber Insulating Gloves • ASTM D1051, Standard Specification for Rubber Insulating Sleeves • ASTM F496, Standard Specification for In-Service Care of Insulating Gloves and Sleeves • ASTM F696, Standard Specification for Leather Protectors for Rubber Insulating Gloves and Mittens
Rainwear	• ASTM F1891, Standard Specification for Arc- and Flame-Resistant Rainwear
Face protective products	• ASTM F2178, Standard Test Method for Determining the Arc Rating of Face Protective Products

NOTE: ASTM—American Society for Testing and Materials

TYPES OF PROTECTION

- Although the product may pass the generally accepted vertical flame test, the characteristic is temporary. Each time the garment is laundered, the product's ability to resist ignition is reduced. The garment becomes less protective with each laundering until the protective characteristic essentially disappears.

- Treating a meltable fabric with a flame-retardant chemical may cause the fabric to resist flame spread, but such treatment does not change the inherent melting property.

An incident energy rating of FR clothing that is determined by testing in accordance with *ASTM F1959, Standard Test Method for Determining the Arc Rating of Materials for Clothing*, means that the clothing will prevent a second-degree burn 50 percent of the time at the assigned rating.

- To ensure avoiding a second-degree burn, the worker must wear clothing with a higher incident-energy rating.

- Without saying so directly, current consensus standards suggest that the risk associated with the current rating system is acceptable.

- A second-degree burn generally is described as a non-curable burn. The skin tissue may regenerate without scarring.

- To achieve greater assurance of avoiding all burns, consensus requirements for protective equipment must be exceeded.

FR clothing fails in two modes, and the assigned arc rating is based on the failure modes.

- In the first failure mode, the incident energy exceeds the thermal insulating ability of the garment.

- In the second failure mode, the garment chars and breaks open (breakthrough) to expose the surface under the garment.

⌨ TYPES OF PROTECTION

Label Requirements

ASTM F1506 requires the clothing manufacturer to provide certain information on the label in each garment **(see Figure 32)**. The information is intended to provide a worker with sufficient information to properly select and use FR protection. The label must contain at least the following information:

- A tracking code

- Indication that the garment conforms to *ASTM F1506*

- Manufacturer's name

- Size and associated standard label information.

- Care instructions and fiber content

- Arc rating, either arc-thermal performance value (ATPV) or energy breakthrough (E_{BT})

FIGURE 32   Clothing label.

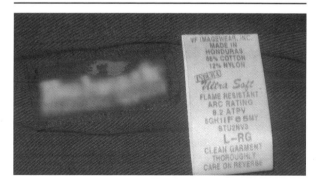

TYPES OF PROTECTION

A worker must determine that each item of FR clothing is suitable for the expected exposure by comparing arc rating on the clothing label with the incident energy exposure level identified in the hazard/risk analysis.

- When a worker performs the hazard/risk analysis required by *NFPA 70E*, he or she will know the incident energy level of potential exposure.

- By reading the label in the FR clothing, the worker knows what level of thermal protection can be expected from the FR clothing.

- If the arc rating of the equipment listed on the clothing label exceeds the expected exposure, the FR clothing is acceptable for that work task.

Thermal Barrier

- Electrical shock protection in the form of rated rubber products reduces the risk of electrical shock or electrocution by reducing the amount of current that might flow in a worker's body.

- The rubber products introduce an additional barrier that resists the flow of electrical current.

- Likewise, FR clothing reduces the risk of thermal burn by reducing the amount of thermal energy that might flow onto a worker's body.

- The FR clothing introduces an additional barrier, which resists the flow of thermal energy.

Selection and Use

The most severe injuries from exposure to an electrical arc are the result of flammable or meltable clothing igniting. In general, the duration of an electrical arc is limited by the clearing time of the overcurrent device. Consensus standards and codes guide the selection of overcurrent devices.

⚡ TYPES OF PROTECTION

- Overcurrent devices function by monitoring the amount of current flowing in the circuit; when the amount of current exceeds the trip setting of the overcurrent device, the circuit is opened, and the source of energy is removed.

- Contemporary codes and standards permit overcurrent devices to be large to reduce the chance of nuisance trips.

- Architects and engineers generally specify circuit breakers and fuses that will reduce the risk of fire in the building and protect the equipment from destruction should a fault occur. Fire was one of the first hazards associated with the use of electrical energy and remains a primary concern.

Electrical equipment usually is constructed from metal components such as copper or aluminum conductors, electrical insulating components, and steel structures or enclosures.

- The metal components of electrical equipment remain solid until the temperature is elevated to the melting point.

- A slightly higher temperature results in the liquid metal beginning to evaporate by boiling, becoming metal vapor.

- The change of state of the metal components begins when the temperature is in the range of 1800°F. Within a few hundred degrees, the metal components evaporate and become metal vapor.

- Human tissue begins to be destroyed when the temperature is elevated to about 150°F and held at that temperature for one second.

- Cells are destroyed in one-tenth of one second when the temperature of the tissue is elevated to about 200°F.

- To avoid a burn injury, then, both the temperature and the duration of the exposure must be limited **(see Table 5)**.

 **TYPES OF PROTECTION**

TABLE 5   Effect of Temperature on Human Skin

Skin Temperature	Time to Reach Temperature	Damage Caused
110ºF	6.0 hours	Cell breakdown begins
158ºF	1.0 second	Total cell destruction
176ºF	0.1 second	Second-degree burn
200ºF	0.1 second	Third-degree burn

Source: Jones, Ray A. and Jane G. Jones, *Electrical Safety in the Workplace*, Sudbury, MA: Jones and Bartlett Publishers, 2000

Limiting Fault-Current Time

- Overcurrent devices remove the source of energy from the arc in a predetermined period of time. Overcurrent devices do not clear the arcing fault instantaneously.

- When a fuse element begins to melt, current continues to flow until the opening in the melting element is large enough to break the circuit.

- Fuses contain a substance similar to sand, which flows into the opening in the melting element and quenches the arc.

- If the fuse is a current-limiting fuse and the fault current is large, the element begins to melt in one-quarter of one cycle, or about 4.5 milliseconds, and clears the fault in less than two cycles, or 34 milliseconds.

- If the current in the arcing fault is below the current-limiting range of the fuse, however, the arcing fault might continue for several seconds.

◨ TYPES OF PROTECTION

Protective Clothing Requirements

To avoid injury, a worker must select protective clothing that has the following characteristics:

- The clothing must not ignite or continue to burn after the arcing fault has been removed.

- The clothing must not melt onto or into the worker's skin as a result of being exposed to the arc.

- The clothing must provide sufficient thermal insulation to prevent the worker's skin tissue from being heated to destruction.

The Flash Protection Boundary

One primary purpose of an arc flash analysis is to determine the flash protection boundary.

- The label on the equipment should identify the flash protection boundary, either in meters or feet and inches.

- Workers should determine the flash protection boundary from the label on the equipment. The flash protection boundary of an idealized arc in free air is a spherical shape, measured from the potential arc to any part of a worker's body **(see Figure 33)**.

- When any part of a worker's body is close to the source of a potential arc, a thermal burn injury is possible and the worker's body must be protected from a possible burn.

TYPES OF PROTECTION

FIGURE 33 Arc-flash protection boundary.

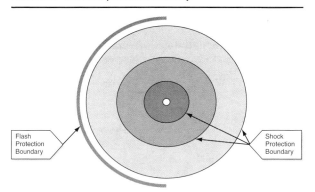

Flash Protection Boundary

Shock Protection Boundary

Reprinted with permission from NFPA 70E®, *Handbook for Electrical Safety in the Workplace,* Copyright © 2004, National Fire Protection Association. This reprinted material is not the complete and official position of the NFPA on the referenced subject, which is represented only by the standard in its entirety.

◻ OTHER EXPOSURE CONSIDERATIONS

- When FR clothing is necessary for a work task, all buttons, zippers, and other closing mechanisms must be completely closed.

- The top fastener near the worker's neck must be closed to minimize the chance that high temperature gasses could get behind the clothing and breach the protective characteristic.

- *ASTM F1506* ensures that adequate layers of FR material protect fasteners on all FR-rated clothing from the extreme heat that could cause a worker to be burned.

- The outer layer of the worker's protective clothing must be flame resistant.

- Any label or identifying patch must be made from the same protective material as the clothing.

- Any garment worn beneath the protective clothing should be made from flame-resistant fabric but must not be made from meltable or easily ignitable fabric.

- However, if the protective clothing will prevent an injury, it also will protect cotton underwear from igniting.

- The head and face of a worker who must perform a task within the flash protection boundary are exposed to the potential arc flash.

- If the worker's head is within the flash protection boundary, he or she should select and wear protective equipment that will eliminate or minimize the risk of exposure to air or gasses that are very hot. Protective equipment that covers the head entirely is available.

- Face shields and balaclava are also available. Note, however, that a balaclava also requires face protection **(see Figure 34)**.

- Equipment intended to protect a worker's head and face is assigned an arc rating that mirrors the arc rating assigned to FR clothing.

OTHER EXPOSURE CONSIDERATIONS

FIGURE 34   Balaclava (head sock). (Courtesy of PGI, Inc., King Cobra™, Ultimate Style Hood—ATPV Rating 23.3)

- When a worker performs the necessary hazard/risk analysis, he or she must consider the location of the work task in relation to his or her body.

- If the work task is near the floor, the worker must consider that hot air and gasses could get behind the protective equipment at the lower extremity.

- Injuries to a worker's legs are possible from the entrance of hot air and gasses under the leg protection on the clothing.

- The worker should ensure that protection for the bottom lower extremities prevents hot air and gasses from getting behind it.

- If the work task (and potential arcing fault) is at an elevated position, the risk of hot air and gasses getting behind the protective equipment near the worker's feet is reduced.

 **PROTECTIVE CHARACTERISTICS**

Testing Flame-Resistant Fabrics

- Flame-resistant fabrics are tested as defined by *ASTM F1506*.

- Representative samples of the fabrics are subjected to the vertical flame test described in *ASTM D6413, Standard Test Method for Flame Resistance of Textiles*.

- This test determines that the fabric will not ignite and burn.

- The material is permitted to continue to flame for a period not exceeding 2 seconds after the test flame is removed and 5 seconds after an arc test.

- The char length in the vertical flame test is not permitted to exceed 6 inches.

- The test is conducted on new fabric and repeated on fabric that has been laundered 25 times.

⚡ ARC RATING

- After the flammability of the fabric has been established, the fabric is then subjected to an arc test as described in *ASTM F1959*.

- An arc rating is established for each fabric sample when the F1959 test is conducted. The after flame of the fabric must not exceed 5 seconds.

- Some fabrics exhibit an "afterglow" that exceeds 5 seconds. However, the afterglow is a property of the fabric and is not considered a flame.

- The product cannot exhibit any indication of melting or dripping when either the arc test or the vertical flame test is conducted.

- The arc rating determines the protective characteristics of the fabric.

- When the product is sold to protect workers from arcing faults, clothing manufacturers are required to provide the arc rating on the label.

- Clothing manufacturers also are permitted to indicate the arc rating on the clothing surface, such as a shirt pocket or sleeve.

- National consensus standards assign protective categories for arc ratings based on the protective nature of the equipment. **Table 6** illustrates the expected arc rating for various clothing categories.

- The protective nature of clothing categories applies to all arc-rated equipment. (The clothing description in **Table 7** is for illustration purposes only.)

- Clothing manufacturers typically assign categories to various clothing combinations.

 **ARC RATING**

TABLE 6   Table 130.7(C)(10)   Protective Clothing and Personal Protective Equipment (PPE)

Hazard/Risk Category	Protective Clothing and PPE
Hazard/Risk Category 0	
Protective Clothing, Nonmelting (according to ASTM F 1506-00) or Untreated Natural Fiber	Shirt (long sleeve) Pants (long)
FR Protective Equipment	Safety glasses or safety goggles (SR) Hearing protection (ear canal inserts) Leather gloves (AN) (Note 2)
Hazard/Risk Category 1	
FR Clothing, Minimum Arc Rating of 4 (Note 1)	Arc-rated long-sleeve shirt (Note 3) Arc-rated pants (Note 3) Arc-rated coverall (Note 4) Arc-rated face shield or arc flash suit hood (Note 7) Arc-rated jacket, parka, or rainwear (AN)
FR Protective Equipment	Hard hat Safety glasses or safety goggles (SR) Hearing protection (ear canal inserts) Leather gloves (Note 2) Leather work shoes (AN)
Hazard/Risk Category 2	
FR Clothing, Minimum Arc Rating of 8 (Note 1)	Arc-rated long-sleeve shirt (Note 5) Arc-rated pants (Note 5) Arc-rated coverall (Note 6) Arc-rated face shield or arc flash suit hood (Note 7) Arc rated jacket, parka, or rainwear (AN)
FR Protective Equipment	Hard hat Safety glasses or safety goggles (SR) Hearing protection (ear canal inserts) Leather gloves (Note 2) Leather work shoes

 **ARC RATING**

TABLE 6 Protective Clothing and Personal Protective Equipment (PPE)—continued

Hazard/Risk Category	Protective Clothing and PPE
Hazard/Risk Category 2*	
FR Clothing, Minimum Arc Rating of 8 (Note 1)	Arc-rated long-sleeve shirt (Note 5) Arc-rated pants (Note 5) Arc-rated coverall (Note 6) Arc-rated arc flash suit hood (Note 10) Arc-rated jacket, parka, or rainwear (AN)
FR Protective Equipment	Hard hat Safety glasses or safety goggles (SR) Hearing protection (ear canal inserts) Leather gloves (Note 2) Leather work shoes
Hazard/Risk Category 3	
FR Clothing, Minimum Arc Rating of 25 (Note 1)	Arc-rated long-sleeve shirt (AR) (Note 8) Arc-rated pants (AR) (Note 8) Arc-rated coverall (AR) (Note 8) Arc-rated arc flash suit jacket (AR) (Note 8) Arc-rated arc flash suit pants (AR) (Note 8) Arc-rated arc flash suit hood (Note 8) Arc-rated jacket, parka, or rainwear (AN)
FR Protective Equipment	Hard hat FR hard hat liner (AR) Safety glasses or safety goggles (SR) Hearing protection (ear canal inserts) Arc-rated gloves (Note 2) Leather work shoes
Hazard/Risk Category 4	
FR Clothing, Minimum Arc Rating of 40 (Note 1)	Arc-rated long-sleeve shirt (AR) (Note 9) Arc-rated pants (AR) (Note 9) Arc-rated coverall (AR) (Note 9) Arc-rated arc flash suit jacket (AR) (Note 9) Arc-rated arc flash suit pants (AR) (Note 9) Arc-rated arc flash suit hood (Note 9) Arc-rated jacket, parka, or rainwear (AN)

(continues)

 **ARC RATING**

TABLE 6   Protective Clothing and Personal Protective Equipment (PPE)—continued

Hazard/Risk Category	Protective Clothing and PPE
FR Protective Equipment	Hard hat FR hard hat liner (AR) Safety glasses or safety goggles (SR) Hearing protection (ear canal inserts) Arc-rated gloves (Note 2) Leather work shoes

AN = As needed (optional); AR = As required; SR = Selection required
NOTES:

1. *See Table 130.7(C)(11). Arc rating for a garment or system of garments is expressed in cal/cm2.*
2. *If rubber insulating gloves with leather protectors are required by Table 130.7(C)(9), additional leather or arc-rated gloves are not required. The combination of rubber insulating gloves with leather protectors satisfies the arc flash protection requirement.*
3. *The FR shirt and pants used for Hazard/ Risk Category 1 shall have a minimum arc rating of 4.*
4. *Alternate is to use FR coveralls (minimum arc rating of 4) instead of FR shirt and FR pants.*
5. *FR shirt and FR pants used for Hazard/ Risk Category 2 shall have a minimum arc rating of 8.*
6. *Alternate is to use FR coveralls (minimum arc rating of 8) instead of FR shirt and FR pants.*
7. *A face shield with a minimum arc rating of 4 for Hazard/Risk Category 1 or a minimum arc rating of 8 for Hazard/Risk Category 2, with wrap-around guarding to protect not only the face, but also the forehead, ears, and neck (or, alternatively, an arc-rated arc flash suit hood), is required.*
8. *An alternate is to use a total FR clothing system and hood, which shall have a minimum arc rating of 25 for Hazard/Risk Category 3.*
9. *The total clothing system consisting of FR shirt and pants and/or FR coveralls and/or arc flash coat and pants and hood shall have a minimum arc rating of 40 for Hazard/Risk Category 4.*
10. *Alternate is to use a face shield with a minimum arc rating of 8 and a balaclava (sock hood) with a minimum arc rating of 8 and which covers the face, head and neck except for the eye and nose areas.*

 **ARC RATING**

TABLE 7   Table 130.7(C)(11)   Protective Clothing Characteristics
Typical Protective Clothing Systems

Hazard/Risk Category	Clothing Description	Required Minimum Arc Rating of PPE [J/cm²(cal/cm²)]
0	Nonmelting, flammable materials (i.e., untreated cotton, wool, rayon, or silk, or blends of these materials) with a fabric weight at least 4.5 oz/yd²	N/A
1	Arc-rated FR shirt and FR pants or FR coverall	16.74 (4)
2	Arc-rated FR shirt and FR pants or FR coverall	33.47 (8)
3	Arc-rated FR shirt and pants or FR coverall, and arc flash suit selected so that the system arc rating meets the required minimum	104.6 (25)
4	Arc-rated FR shirt and pants or FR coverall, and arc flash suit selected so that the system arc rating meets the required minimum	167.36 (40)

NOTE: Arc rating is defined in Article 100 and can be either ATPV or E_{BT}. ATPV is defined in ASTM F 1959, Standard Test Method for Determining the Arc Thermal Performance Value of Materials for Clothing, as the incident energy on a material or a multilayer system of materials that results in a 50% probability that sufficient heat transfer through the tested specimen is predicted to cause the onset of a second-degree skin burn injury based on the Stoll curve, cal/cm². E_{BT} is defined in ASTM F 1959 as the incident energy on a material or material system that results in a 50% probability of breakopen. Arc rating is reported as either ATPV or E_{BT}, whichever is the lower value.

 **ARC RATING**

ASTM F1506 is a performance specification that identifies specific tests and acceptable results.

- The specification covers fabrics that are woven or knitted. The specification is not intended for use with other fabric construction.

- In general, fabric is measured in weight per unit of area.

- Fabric produced for use as FR clothing is also measured in ounces per square yard. As the weight per square yard of FR fabric increases, the thermal insulating ability also increases.

- In most cases, fabric from different clothing manufacturers is constructed from different textile products and blends of products from different textile manufacturers.

- The weight of the FR protective clothing is not an accurate indicator of the degree of protection provided by the clothing. The arc rating is the only reasonable indicator of the protection provided by the clothing.

- Employers have a common practice of providing patches with an employee's name or a company logo. Employees are then encouraged to attach their name and company logo to their shirt or uniform.

- Any patch or logo attached to FR clothing also must be flame resistant. Patches that are flammable or meltable provide fuel that can ignite or melt and increase the effect of an arcing fault.

- The protective characteristic of FR clothing depends on the surface of the fabric being clean and free from flammable material such as grease and oil.

ARC RATING

- Wiping cloths should not be kept on a worker's body or clothing pocket when he or she is at risk of being exposed to a potential arcing fault. Just as a flammable patch adds fuel to the fire, grease and oil also add fuel to the fire.

- Wiping cloths that are kept in a pocket are likely to ignite and burn if exposed to an arcing fault. A burning wiping cloth in a pocket is likely to overcome the protection provided by the FR clothing.

# INSULATING FACTOR OF LAYERS

Air is a relatively good thermal insulator.

- If two or more layers of FR fabric are used as part of a protective system, a layer of air is trapped between the layers.

- The air layer provides additional thermal insulation between a worker's skin and any potential arcing fault. Therefore, two layers of FR fabric provide greater protection than the arithmetic sum of the protective rating of the individual layers. (See Table 7, *NFPA 70E* Table 130.7(C)(11), for PPE Hazard Risk Categories and Required Minimum Hazard/Risk Arc Rating.)

The basic idea of providing protection is adding thermal insulation between a worker and an arcing fault. As indicated above, a layer of air increases the thermal protection. FR clothing, then, should be loose fitting but not so loose as to interfere with the worker's movement. When the FR clothing is in contact with the worker's skin, thermal energy can be conducted through the fabric to the worker.

- Fabric used to construct flame-resistant clothing provides only thermal insulation. It provides no protection from shock or electrocution. However, FR-protective equipment does not increase the chance of electrical shock unless the material is metalized or contains a conductive component, such as carbon. Although the reflectivity of a metalized garment would reduce the amount of energy conducted through the garment, the conductive garment would increase the chance of initiating an arcing fault. Some protective clothing contains a small amount of carbon to reduce the effect of static electricity. Small amounts of carbon embedded in the thread do not increase exposure to shock. Users should contact the clothing or fabric manufacturer for further details.

- An FR-rated switching hood provides greater protection than a face shield. In addition to the thermal protection provided by an FR-rated hood, protection is also provided by a small quantity of air trapped in the hood. Should an arcing fault occur, the uncontaminated air inside the hood might be an advantage for a worker **(see Figure 35)**.

 **INSULATING FACTOR OF LAYERS**

FIGURE 35   FR-rated switching hood. (© 2009, VF Imagewear, Inc. Used with permission.)

- A face shield provides protection from energy that is transmitted in the form of radiation. Energy that is transmitted by convection could flow behind the face shield and cause a burn. A face shield provides protection from flying parts and pieces that may be expelled by the blast component of the arcing fault.

 **PPE CONFIGURATIONS**

PPE can consist of many different configurations, including various combinations of flame-resistant clothing and protective equipment **(see Figures 36–40)**.

- A worker might choose to wear coveralls or a shirt-and-pants combination.

- The PPE might consist of cotton clothing.

FIGURE 36   FR shirt and pants. (Courtesy of Applied Concepts and Z-Studios, Ventura, CA)

 PPE CONFIGURATIONS

FIGURE 37   FR long coat. (Courtesy of Applied Concepts and Z-Studios, Ventura, CA)

FIGURE 38   Leather gloves. (Courtesy of Ironwear)

PPE CONFIGURATIONS

FIGURE 39   Complete flash suit. (Courtesy of Salisbury Electrical Safety LLC)

- The PPE might consist of a single layer, or it might consist of two or more layers.

- The PPE might consist of both natural fiber material and FR material. The manufacturer should be consulted to determine the overall rating of any multiple-layer protective system.

Workers must not wear clothing made from fabrics that melt, such as polyester, nylon, acetate, and similar products. Clothing containing blends of these products must not be worn unless the product is assigned an arc rating by the manufacturer.

 **PPE CONFIGURATIONS**

FIGURE 40   FR workclothes. (Courtesy of Applied Concepts and Z-Studios, Ventura, CA)

Many undergarments are made of polyester, nylon, acetate, and similar products. If an undergarment melts, the melted fabric will destroy any skin tissue that it contacts. Workers who wear FR-protective equipment should wear undergarments of cotton or flame-resistant fabric only.

AN ARCING FAULT CONVERTS ELECTRICAL ENERGY INTO OTHER FORMS OF ENERGY

Flame-resistant products discussed in this section provide protection from the energy that is converted into thermal energy. All electrical arcs also convert some energy into pressure. The pressure wave created when an arcing fault is initiated produces a wave that results in a force known as blast.

Flame-resistant PPE offers little protection from forces associated with the pressure wave. As the electrical energy in the arc increases, both the thermal and blast forces increase. Arcing faults that produce incident energy greater than 40 calories per square inch produce pressure waves that may cause injury to a worker, regardless of the amount of FR protective equipment the worker is wearing. If the flash hazard analysis indicates that the incident energy is 40 calories per square inch or more, the work task should not be performed while the circuit or equipment is energized.

CLOTHING DESIGNED ESPECIALLY FOR WOMEN

Until a few years ago, FR clothing was made in sizes for male workers only, although women might wear it. Several manufacturers such as ArcWear™, Bulwark®, Tyndale, and Workrite® now offer FR and arc-rated clothing, including both outerwear and underwear, in standard sizes for women **(see Figure 41 and Figure 42)**.

Men can wear cotton underwear under their arc-rated clothing with no danger of it melting. However, some women's underwear is made from polyester or similar materials that could melt in an arc-flash event, and some bras contain metal underwires, hooks, or clips that should not be worn (even under FR clothing) if there is potential for exposure to the arc-flash hazard.

FIGURE 41   Arc-rated bra. (Courtesy of Hugh Hoagland, ArcStore.com)

FIGURE 42   FR clothing for women. (Courtesy of Applied Concepts and Z-Studios, Ventura, CA)

⎍ PROTECTIVE MATERIALS

- Leather products are intended to avoid abrasion and penetration.

- Although leather provides sound protection from abrasion and from penetration by parts that are static, moving parts and pieces ejected by an arc blast will be impeded only partially. Some FR clothing contains material (such as Kevlar®) that normally is used to stop small flying objects such as bullets.

- If the FR clothing successfully stops the moving part, the energy contained in the moving object will be transferred to the person. Face shields, safety glasses (spectacles), and viewing windows in hoods are intended to provide protection from impact. However, these products are not tested in an environment where objects of significant size and momentum might exist.

- Rubber products worn for protection from electrical shock provide a layer that can resist penetration and abrasion. Again, however, energy contained in the flying parts and pieces is transferred to the person, even if the rubber protection is not penetrated.

The bottom line is that no products exist that can protect a person from the blast effects of an arcing fault. Products worn for protection from other electrical hazards can provide some resistance to penetration, but they do not absorb the kinetic energy in the momentum of flying parts and pieces. Only two options exist to protect from a potential arc blast:

1. The equipment must be placed in an electrically safe work environment.

2. There must be sufficient distance from the source of the potential arc.

 **PROTECTIVE MATERIALS**

NFPA 70E suggests that a safe electrical work environment has three components:

- *NEC*-compliant installation

- Proper maintenance

- Safe work practices

Adequate maintenance of electrical systems improves reliability and reduces employee exposure to electrical hazards. The following examples illustrate this axiom:

- When overcurrent devices (circuit breakers and fuses) function as designed to clear faults quickly, arc-flash and arc-blast hazards are reduced.

- When grounding and bonding systems function properly, shock hazard and the sensation of "touch voltage" are reduced for all employees in the workplace.

QUALIFIED PERSONS

Only qualified persons are permitted to perform maintenance on electrical equipment, circuits, and installations.

GENERAL ELECTRICAL EQUIPMENT

Chapter 2 of *NFPA 70E* contains general requirements for maintaining electrical equipment. It refers to *NFPA 70B, Recommended Practice for Electrical Equipment Maintenance*, for specific maintenance methods and tests. Chapter 2 is divided into 10 articles.

Article 205—General Maintenance Requirements

Article 210—Substations, Switchgear Assemblies, Switchboards, Paperboards, Motor Control Centers, and Disconnect Switches

Article 215—Premises Wiring

Article 220—Controller Equipment

Article 225—Fuses and Circuit Breakers

Article 230—Rotating Equipment

Article 235—Hazardous (Classified) Locations

Article 240—Batteries and Battery Rooms

Article 245—Portable Electric Tools and Equipment

Article 250—Personal Safety and Protective Equipment

Chapter 2 of *NFPA 70E* summarizes maintenance requirements for electrical installations. Qualified persons should follow the latest edition of *NFPA 70B-2006, Recommended Practice for Electrical Equipment Maintenance*, when maintaining electrical equipment and systems.

⚡ SPECIAL ELECTRICAL EQUIPMENT

NFPA 70E also includes safety-related work practices for installing and maintaining four types of special electrical equipment. Electricians and electrical contractors doing typical residential-commercial-industrial jobs rarely work on these specialized types of equipment. *NFPA 70E*, Chapter 3, covers the following:

1. Electrolytic cells

2. Batteries and battery rooms

3. Lasers

4. Power electronic equipment

5. Research and development laboratories

SPECIAL ELECTRICAL EQUIPMENT

1. Electrolytic Cells (Article 310)

Electrolytic cells are industrial equipment used in the production of metals such as copper, aluminum, and zinc. Each cell (or pot) is a DC electric furnace, and multiple DC furnaces connected in series are called a *cell line*. The working area surrounding this equipment is called a *cell-line working zone*.

Heat Causes Special Hazards

Temperatures inside electrolytic cells can reach 900°C, and operating voltages can reach 1000 volts or more. This combination can make conventional PPE (such as flash suits) impractical. Using regular *NFPA 70E* rules to protect qualified persons against available arc-flash incident energy can incapacitate them through overheating.

Special Safety Techniques

For this reason, *NFPA 70E*, Article 310, permits employers to perform a flash hazard analysis and determine what training, safe work practices, and PPE are required for each task inside a cell-line working zone.

# SPECIAL ELECTRICAL EQUIPMENT

2. Batteries and Battery Rooms (Article 320)

Storage batteries are commonly used with uninterruptible power supplies (UPS), some types of power electronic equipment, and emergency control power for switching. The scope of this article states that it applies to batteries and battery rooms operating at 50 to 650 volts and a capacity exceeding 1 kWh. However, *NFPA 70E* Article 320.7 requires that warning signs be posted on systems operating at 50 volts or more, and *NFPA 70E* Article 320.4 states that permanently installed batteries operating at more than 24 volts and 10 ampere-hours must be in a battery room or enclosure.

Safety Requirements

- *NFPA 70E*, Article 320, deals with construction of battery rooms, working clearances between battery racks, and ventilation to prevent hazardous buildups of flammable hydrogen.

- It also requires workers to use nonsparking and insulated tools when working on batteries and specifies PPE needed to work with batteries.

- Another important item is an emergency shower for rinsing eyes or skin in case of electrolyte (battery acid) spills.

# SPECIAL ELECTRICAL EQUIPMENT

3. Lasers (Article 330)

This brief article specifies what employers must do with respect to:

- employee training

- eye protection

- warning signs

- engineering controls

- employee safety responsibilities

Unfortunately, the article stops there and does not go into describing the different laser classifications. Lasers are classified according to their capabilities of producing injury to personnel. These classifications range from Class I (no injury) to Class IV (able to cut thick steel). Class I lasers include levelers used on construction sites and laser pointers.

 **SPECIAL ELECTRICAL EQUIPMENT**

4. Power Electronic Equipment (Article 340)

This article discusses a range of industrial, medical, and communications gear, including:

- electric arc welding equipment

- radio, radar, and television antennas

- dielectric and radio frequency induction equipment

- diathermy devices

- equipment that uses rectifiers and inverters, such as uninterruptible power supplies and frequency converters

Diathermy devices are medical equipment that use high frequency radiation, microwaves, or ultrasound to heat and destroy abnormal cells.

SPECIAL ELECTRICAL EQUIPMENT

5. Research and Development Laboratories (Article 350)

This article discusses equipment and techniques that are acceptable for use in laboratory conditions. The article establishes exceptions from Chapter 1 requirements that apply to tasks that are acceptable and necessary for special equipment used in research and development facilities but might be unacceptable in routine industrial conditions.

 **SAFETY TECHNIQUES FOR USERS**

NFPA 70E, Article 340, contains little specific information for electrical construction and maintenance personnel. It primarily provides background information on hazards associated with power electronic equipment such as the following:

- Ensuring that equipment is installed properly

- Identifying and guarding dangerous equipment

- Providing adequate training and instructions

- Maintaining a clean and clear work area

- Providing adequate illumination of the work area

Chapter 3 of *NFPA 70E* summarizes safe working practice requirements for working on energized special electrical equipment.

O ENFORCES COMPLIANCE WITH *NFPA 70E*?

Unlike the *NEC*, states, cities, and counties do not adopt *NFPA 70E* for regulatory purposes. State and municipal electrical inspectors do not enforce *NFPA 70E* unless the standard is adopted by local ordinance.

Instead, facility owners often require that both their own employees and outside contractors comply with *NFPA 70E* while doing electrical construction and maintenance work. Increasingly, owners are requiring that contractors working for them provide evidence that their crews have been trained in *NFPA 70E* safety practices. Doing so reduces the customers' liability exposure and insurance premiums.

HOW IS *NFPA 70E* RELATED TO THE *NATIONAL ELECTRICAL CODE*®?

The *National Electrical Code* describes design and installation of premise wiring systems, while *NFPA 70E* describes how to perform the work safely. Said another way, the *NEC* applies to equipment, circuits, and devices, while *NFPA 70E* applies to people. Both publications are also closely related to a third industry standard: *NFPA 70B, Recommended Practice for Electrical Maintenance.* Here is how the three work together:

- *National Electrical Code* (*NFPA 70*)—Describes how to design and install electrical systems that operate safely. It deals with subjects like overcurrent protection, conductor ampacity, wiring methods, equipment ratings, and grounding. The *NEC* does not cover maintenance of electrical equipment, or safe work practice issues such as when workers should use PPE.

- *NFPA 70E*—Describes how to perform installation and maintenance work safely. It covers safe electrical work practices such as lockout–tagout, and when workers should use insulated tools and wear FR clothing.

- *NFPA 70B*—Describes maintenance practices that keep electrical systems running reliably and safely. It does not cover design or installation issues and does not discuss safe work practices.

Knowledge Is Critical

Good electrical knowledge is needed to use *NFPA 70E* safely and effectively. *NFPA 70E* is organized in a way that is similar to the *NEC*, which helps electrical workers who are already familiar with the *NEC* to understand and apply the safety standard.

HOW IS *NFPA 70E* RELATED TO OSHA REGULATIONS?

All employers, in all industries, are legally required to follow Occupational Safety and Health Administration (OSHA) regulations to protect their workers from job-related hazards. *NFPA 70E* is an American National Standard, developed by the publisher of the *National Electrical Code*; it parallels OSHA's electrical safety regulations.

However, *NFPA 70E* is not the *same* as OSHA regulations. It is more up to date than the OSHA electrical safety regulations. This guide summarizes the electrical safe work practices defined in *NFPA 70E*. For more information about how the standard is related to OSHA electrical regulations, see the Appendix and the following publications:

- *NFPA 70E Handbook for Electrical Safety in the Workplace*

- *Stallcup's OSHA Electrical Regulations Simplified* (Jones and Bartlett Publishers)

NFPA 70E is a private-sector, voluntary American National Standard that parallels the following OSHA safety regulations:

- OSHA 29 CFR 1910, General Industry Standards, Subpart S— Electrical (covers maintenance and repairs on existing systems).

- OSHA 29 CFR 1926, Construction Industry Standards, Subpart K— Electrical (covers new construction).

Most electrical contractors are required to follow both sets of OSHA safety regulations, depending on the type of work they are performing; 29 CFR 1926 covers *new* construction, before the facility is energized, while 29 CFR 1910 covers *maintenance* work. The two standards are similar but not quite matching, which can cause confusion.

OSHA Involvement

NFPA 70E was developed originally at OSHA's request. Because the federal government rulemaking process is slow and cumbersome, keeping

HOW IS *NFPA 70E* RELATED TO OSHA REGULATIONS?

OSHA regulations up to date with current technology and work practices is difficult. For example, the OSHA 29 CFR 1926, Subpart K, regulations still reference the 1984 *National Electrical Code*.

For this reason, private industry uses the *NFPA 70E* standard to "lead" the two sets of OSHA workplace safety regulations that govern electrical work: 1910 Subpart S for general industry applications, and 1926 Subpart K for the construction industry. OSHA personnel participate in the *NFPA 70E* Technical Committee.

NFPA 70E Enforcement

The practical result of complying with the safe work practices defined in *NFPA 70E* is in most cases complying also with the applicable OSHA regulations. While OSHA safety and health compliance officers do not enforce *NFPA 70E per se*, there is a growing tendency for them to rely on *NFPA 70E* under the so-called "general duty" clause. Section 5 (a)(1) of the Occupational Safety and Health Act requires employers to furnish safe workplaces that are free from "[r]ecognized hazards . . . likely to cause death or serious harm to employees."

⚡ WHO IS RESPONSIBLE FOR ELECTRICAL SAFETY?

NFPA 70E states that both employers and employees are responsible for preventing injury.

- Employers are responsible for establishing safe work practices and for training employees in those practices.

- Employees are responsible for following the safe work practices established by their employer.

- Multiple employers often work together on the same construction site or in buildings and similar facilities. Some might be onsite personnel working for the host employer, while others are "outside" personnel such as electrical contractors, mechanical and plumbing contractors, painters, or cleaning crews.

- *NFPA 70E* requires that when an onsite employer and outside employer(s) work together on or near electrical installations, they must coordinate their safety procedures.

- This coordination must include a meeting and documentation such as notes or minutes to record what is discussed and decided.

- Outside contractors often are required to follow the host employer's safety procedures.

- Multiple employers involved in the same project sometimes decide to follow the most stringent set of safety procedures.

- Whichever approach is taken, the decision should be recorded in the safety meeting documentation.

Who Is Responsible for PPE?

Both *NFPA 70E* and OSHA rules require various kinds of PPE, including insulated tools, face shields, and flame-resistant clothing to protect electrical workers. In many cases, employers are required to supply PPE if their electrical safety program requires specific equipment. However, workers are responsible for ensuring that appropriate PPE is used when performing a work task.

WHAT ARE ELECTRICAL HAZARDS?

Historically, shock and electrocution were seen as the primary electrical hazards to people, along with fires of electrical origin. Today, however, awareness of two other electrical hazards is growing: arc flash and arc blast. *NFPA 70E* defines three types of electrical hazards:

1   Electric shock

2. Arc flash

3. Arc blast

Electric Shock Hazard

- Electric shock is the leading cause of deaths due to electricity.

- Several thousand nonfatal electrical shock accidents and several hundred electrocutions (deaths from electric shock) occur each year.

- More than half of the fatalities occur when people are working on conductors and equipment energized at less than 600 volts.

- Shocks happen when a person contacts an exposed energized electrical conductor or circuit part or in some other way becomes part of an electrical circuit.

- Even the current needed to light an old-fashioned Christmas tree bulb (7-1/2 watts, 120 volts) is enough to kill a person if it passes across the chest and through the heart **(see Figure 43)**. *Note: Current, not voltage, is the killer.*

Ground-Fault Circuit Interrupter (GFCI) Protection

- *NFPA 70E*, *NEC*, and OSHA require GFCI protection on receptacles used to provide temporary power for construction and maintenance work.

- GFCI circuit breakers and receptacles disconnect the circuit if fault current to ground (that might be traveling through a worker's body) exceeds 4–6 milliamperes (mA).

WHAT ARE ELECTRICAL HAZARDS?

FIGURE 43   Current flow through the human body.

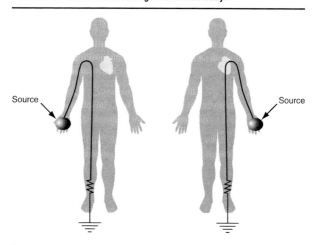

Source

Source

Insulated Gloves and Tools

- Insulated gloves, voltage-rated fiberglass tools, and other insulating PPE are other ways of protecting workers from electric shock hazard.

- They are not intended for protection from arc-flash and arc-blast hazards.

Shock Approach Boundaries

- *NFPA 70E* defines three shock approach boundaries to energized electrical conductors and equipment: limited, restricted, and prohibited.

- These boundaries help protect workers against electric shock hazard. However, they provide no protection from arc-flash and arc-blast hazards.

WHAT ARE ELECTRICAL HAZARDS?

- Approach boundaries around energized electrical systems are explained further in the earlier section of this guide, Safe Electrical Work Practices.

Arc-Flash Hazard

- When electric current passes through air, temperatures can get hot enough to ignite clothing, melt plastic, and burn human skin up to 10 feet away.

- Each year more than 2,000 people are admitted to hospitals with arc-flash burns.

- In addition to burn danger, the very bright light associated with arc-flash events can cause temporary (and sometimes permanent) blindness.

Causes of Arc Flashes

- Arc flashes are usually short events lasting less than a half-second (30 cycles).

- They can be caused by such things as dropping a tool or metal part across two energized busbars, operating malfunctioning or damaged switching equipment (so that an energized blade comes in contact with a metal case), or voltage testing with an improperly rated meter.

- Although an arc flash can be initiated by equipment failure, the vast majority of arc-flash events result from worker action.

Arc-Flash Protection Boundary

- *NFPA 70E* defines an arc-flash protection boundary around energized electrical conductors and equipment.

- The boundary varies in size depending on the energy available in case of a fault.

- The arc-flash protection boundary is frequently larger than the three shock approach boundaries.

WHAT ARE ELECTRICAL HAZARDS?

Arc Flash and Voltage Levels

- Arc-flash hazard is related to the energy available in case of a fault.

- This energy level is related to the available short-circuit current and the clearing time of the circuit's overcurrent device (circuit breaker or fuse).

- Arc flash does not depend on system voltage alone. Under normal circumstances, 277- and 480-volt electrical systems present very serious arc-flash hazards.

Arc-Blast Hazard

- Arc blast is a more severe version of arc flash. At higher available energy levels, explosion hazards—high pressure, flying shrapnel, and loud noise—are added to the thermal and light dangers of arc flashes.

- Extreme temperatures generated by the electric arc cause explosive expansion of both the surrounding air and of metal in the arc path. Copper expands many thousands of times when it vaporizes, creating a deadly spray of molten metal droplets.

Pressure Wave

This rapid expansion creates a pressure wave that can rupture equipment enclosures, knock workers off ladders, spray molten metal, rupture eardrums, and rip apart PPE.

Burn Dangers

- Oddly, workers exposed to arc blasts sometimes suffer fewer burns than those in arc-flash incidents.

- The blast energy can blow workers away from the arc, thus reducing their heat exposure but increasing the risk of serious falls or impact injuries.

⚡ WHAT ARE ELECTRICAL HAZARDS?

Contact Hazards and Non-Contact Hazards

As these explanations show, the three electrical hazards defined in *NFPA 70E* actually fall into two general categories:

1. Contact hazards (shock and electrocution)

2. Non-contact hazards (arc flash and arc blast)

Different techniques are used to protect workers against each category of hazard.

Contact Hazards (Shock and Electrocution)

People get shocked or electrocuted when they touch an energized (live) part and become part of a circuit.

Qualified persons working on energized electrical systems can be protected against shock and electrocution by taking these precautions:

• Wearing insulated gloves and other clothing

• Using insulated tools

• Observing approach boundaries

Unqualified persons are protected against shock and electrocution by the following:

• Good maintenance practices (keeping electrical equipment in safe operating condition, making sure junction boxes have covers, ensuring that non-current-carrying metal parts of electrical equipment are grounded)

• Barriers and warning signs

• Staying outside the limited approach boundary

WHAT ARE ELECTRICAL HAZARDS?

Non-Contact Hazards (Arc Flash and Arc Blast)

- Sometimes people are injured by being near an electrical failure.

- People can be injured by arc flashes and arc blasts even when they are not touching an exposed energized electrical conductor—sometimes when they are several feet away.

Qualified persons working on an energized electrical system can be protected against arc flash and arc blast by taking these precautions:

- Staying outside the arc-flash protection boundary will help protect from all burns. Nevertheless, there is still a possibility of a first degree burn.

- Wearing and using appropriate PPE, including FR clothing, when they must work within the arc flash protection boundary.

ELECTRICAL SAFETY PROGRAM

NFPA 70E requires that every employer have a written electrical safety program for work involving electrical systems or circuits operating at 50 volts or more. Annex E of *NFPA 70E* describes general principles of such programs. Detailed information about how to set up company electrical safety programs is available from various sources, including the following:

- *Electrical Safety Program Book*, by Ken Mastrullo, Ray A. Jones, and Jane G. Jones (Jones and Bartlett Publishers, www.jbpub.com).

  This book describes how to set up a company electrical safety program and includes a CD-ROM with necessary forms and paperwork in PDF format.

- *Electrical Safety in the Workplace*, by Ray A. Jones and Jane G. Jones (Jones and Bartlett Publishers, www.jbpub.com).

- *A User's Guide to Electrical PPE*, by Ray A. Jones and Jane G. Jones (Jones and Bartlett Publishers, www.jbpub.com).

- *Safety Expert System Software*, Version 3.1 (National Electrical Contractors Association, www.necanet.org/store).

  This software system is designed for electrical contractors, but other types of employers also can use it to set up and maintain electrical safety programs. The interactive system has software modules on different subjects that safety directors might use to create and print out an electrical safety program customized for their particular needs.

Employers can use NECA's *Safety Expert System Software* to set up a complete OSHA-compliant safety program. The software contains modules for other safety subjects such as fall protection, ladders and scaffolding, hazardous materials, bloodborne pathogens, motor vehicle use by employees, and many other subjects. It includes a module for keeping records as required by OSHA.

⌁ TRAINING EMPLOYEES IN ELECTRICAL SAFETY

NFPA 70E requires that employers provide electrical safety training for employees who will work on electrical systems or circuits that are or might be energized at 50 volts or more. This means that *qualified persons* who perform electrical construction and maintenance work must receive electrical safety training.

Unqualified persons, such as office workers, sales people, cleaning crews, and workers in other construction trades who do not perform electrical work do not need electrical safety training. However, unqualified persons must understand that they are not qualified to work on electrical equipment or circuits.

Training Methods

The *NFPA 70E* standard permits electrical safety training to be classroom, on the job, or a combination of the two. Many employers use a combination of the following methods of training qualified persons in electrical safe work practices:

- Toolbox talks

- OSHA 10-hour safety course

- OSHA 30-hour safety course

- *NFPA 70E* training course

On-the-Job Training

Apprentices and helpers work under the supervision of a qualified person such as a journeyman electrician, who instructs them in electrical safety techniques by example and practice.

TRAINING EMPLOYEES IN ELECTRICAL SAFETY

Toolbox Talks

Many employers hold brief electrical safety training classes once a week, or more often, usually at the start of a shift. A foreman or instructor usually leads these classes, printed materials are sometimes handed out, and a written record is kept of who attended. Toolbox talks (also called *tailgate talks*) typically focus on a single safety topic, such as grounding, use of GFCIs, use of PPE, arc-blast hazards, safety awareness, or shock approach boundaries.

Sources of electrical safety toolbox talks in paper, CD-ROM, or video format include the following:

- Jones and Bartlett Publishers, LLC, www.jbpub.com

- National Electrical Contractors Association, www.necanet.org/store

- Coastal Training Technologies Corp., www.coastalsafetytraining.com

OSHA 10-Hour Safety Course

Many electricians, technicians, and other qualified persons take the OSHA 10-hour safety course. This course provides an overview of general OSHA safety rules that apply to all construction sites and similar workplaces. It also provides an introduction to OSHA electrical safety rules. Students receive an official OSHA card after completing the course.

Many facility owners require electricians and others working on their premises to have OSHA 10-hour course completion cards. Some electrical contractors also require this for all field employees.

Training vendors, electrician apprenticeship programs, community colleges, and organizations such National Safety Council (NSC) local chapters teach the OSHA 10-hour course. ClickSafety.com, Inc. has an online OSHA 10-hour course that qualifies individuals for the official completion card (www.osha10.com).

⊔ TRAINING EMPLOYEES IN ELECTRICAL SAFETY

OSHA 30-Hour Safety Course

Many electricians, foremen, general foremen, and supervisors take an OSHA 30-hour safety course. This builds on the OSHA 10-hour course by providing an additional 20 hours of instruction in OSHA electrical safety regulations. Students receive an official OSHA card after completing the course.

Courses are available that concentrate on one of the following:

• OSHA Part 1926, Subpart K (construction)

• OSHA Part 1910, Subpart S (general industry)

Students receive an official OSHA card after completing the course.

Many facility owners require electrical supervisory personnel working on their premises to have OSHA 30-hour course completion cards. Some electrical contractors also require this for their foremen and general foremen.

Training vendors, apprenticeship programs, community colleges, and organizations such as local chapters of the National Safety Council (NSC) and American Society of Safety Engineers (ASSE) teach the OSHA 30-hour course. ClickSafety.com, Inc. has an online OSHA 30-hour course that qualifies individuals for the official completion card (www.osha30.com).

NFPA 70E Safety Course

Many vendors now teach *NFPA 70E* courses, including these organizations:

• National Fire Protection Association, www.nfpa.org

• AVO Training Institute, www.avotraining.com

TRAINING EMPLOYEES IN ELECTRICAL SAFETY

Although some facility owners require *NFPA 70E* "certification" for electrical personnel working on their premises, NFPA does not offer such an official certification. However, students who take *NFPA 70E* training courses normally receive a completion certificate or wallet card from the training vendor.

Job Briefing

NFPA 70E requires that before starting each job, the employee in charge must conduct a job briefing with the employees involved. The content of the briefing should be guided by the results of the required hazard/risk analysis. The extent of the briefing depends on the degree of risk involved and whether the work is routine or repetitive.

NFPA 70E has 15 annexes containing information that helps users comply with the standard. Brief summaries follow:

Annex A—Referenced Publications

This annex lists other industry standards referenced in *NFPA 70E*. These publications are considered part of the requirements of the electrical safety standard. Except for the *National Electrical Code*, most of the referenced publications deal with PPE.

Annex B—Informational References

This annex lists other publications mentioned in *NFPA 70E*. These publications are not considered part of the requirements in the standard.

Annex C—Limits of Approach

This annex describes the three approach boundaries for shock protection: limited, restricted, and prohibited. This subject is covered on page 35.

TRAINING EMPLOYEES IN ELECTRICAL SAFETY

Annex D—Sample Calculations for the Arc-Flash Protection Boundary

This annex describes a manual method for calculating available incident energy to determine the arc-flash protection boundary. As explained on page 36 of this *Ugly's* guide, the default arc-flash protection boundary is 4 feet for systems rated at 600 volts and less when the product of the clearing time and available fault current is 100 kA-cycles.

Annex E—Electrical Safety Program

This annex summarizes basic principles of company electrical safety programs.

Annex F—Hazard/Risk Evaluation Procedure

This annex consists of a flow chart for performing hazard/risk evaluations. The intent of the flow chart is to illustrate a series of increasingly discerning questions to determine if or when an electrical hazard exists. Annex F also describes one method of evaluating risk for a type of work.

Annex G—Sample Lockout–Tagout Procedures

This annex contains a sample company lockout–tagout policy. Page 11 of this guide describes the procedure for establishing an electrically safe work condition.

Annex H—Simplified Two-Category, FR Clothing System

This annex describes one approach to selecting FR clothing that does not require using the *NFPA 70E* tables.

TRAINING EMPLOYEES IN ELECTRICAL SAFETY

Annex I—Job Briefing and Planning Checklist

This annex contains a sample checklist for job briefings to plan electrical work. Page 123 of this *Ugly's* guide discusses job briefings.

Annex J—Energized Electrical Work Permit

This annex contains a sample Energized Electrical Work Permit. Page 32 of this *Ugly's* guide discusses an Energized Electrical Work Permit (sometimes called a *Live Work Permit*).

Annex K—General Categories of Electrical Hazards

This annex describes three types of electrical hazards: shock, arc flash, and arc blast. Pages 113–118 of this *Ugly's* guide covers this subject.

Annex L—Typical Application of Safeguards in the Cell Line Working Zone

This annex describes safeguards for working around electrolytic cell lines covered by *NFPA 70E*, Article 310.

Annex M—Layering of Protective Clothing and Total System Arc Rating

This annex discusses issues associated with the protective nature of multiple layers of arc-rated FR clothing.

Annex N—Example Industrial Procedures and Policies for Working Near Overhead Lines and Equipment

This annex provides one example of a procedure for working on exposed overhead conductors on poles or towers.

TRAINING EMPLOYEES IN ELECTRICAL SAFETY

Annex O—Safety-Related Design Requirements

This annex illustrates that many decisions made during the design of a facility have an important bearing on safe work practices.

Conclusion

Several of the annexes to *NFPA 70E*, particularly Annex H, Annex I, and Annex J, contain useful technical information that is critical to correct application of the standard. The remaining annexes provide general illustrations or administrative information.

INSTALLATIONS COVERED BY *NFPA 70E*

For complete information about which electrical installations are covered or not covered by *NFPA 70E*, see Sections 90.1, 110.6(D)(1)(c), 110.7(F), and 130.1 of *NFPA 70E*. The following lists of **covered** and **not covered** installations provide useful examples:

Covered

NFPA 70E applies to electrical installations operating at 50 volts and above, on the customer side of the service point. Examples include the following:

- 120/240-volt, single-phase, 3-wire power systems

- 208Y/120-volt, 3-phase, 4-wire power systems

- 480Y/277-volt, 3-phase, 4-wire power systems

- 480-volt, 3-phase, 3-wire systems

- Class 1 control wiring operating at up to 150 volts (covered by *NEC* Article 725)

- 120-volt, motor-control wiring

- 600Y/347-volt, 3-phase, 4-wire power systems (*Note: While this system voltage is common in Canada, it is rarely used in the U.S.*)

- Privately owned power systems operating at above 600 volts. These include installations such as industrial power systems operating at 4160Y/2400 volts and campus-wide power distribution systems operating at 7200 volts or 13.2 kilovolts.

NOTE: This guide covers only work on systems operating at 600 volts and below. See the NFPA 70E standard itself to learn about safe work practices for high-voltage electrical systems.

█ INSTALLATIONS NOT COVERED BY *NFPA 70E*

Not Covered

NFPA 70E does not apply to electrical installations operating at less than 50 volts. These installations include most signaling, communications, and control installations, along with low-voltage lighting. Examples of systems *not covered* by *NFPA 70E* include the following:

- Telecommunications systems operating at 48 volts

- Fire alarm systems operating at 24 volts and 32 volts

- Class 2 and 3 wiring systems (covered by *NEC* Article 725)

- Remote control systems such as 12-volt lighting control

- Optical fiber communications systems (*Note: Composite cables containing both optical fibers and electric power conductors operating at 50 volts or higher are covered by NFPA 70E.*)

- Low-voltage lighting systems operating at a maximum of 30 volts (covered by *NEC* Article 411)

- Utility generating, transmission, and distribution systems

- DC power systems for railroads, subways, and similar purposes

- Airplanes, watercraft, and automotive vehicles (other than recreational vehicles)

NFPA 70E DEFINITIONS

NOTE: *Some* NFPA 70E *definitions include fine print notes (FPNs), which are shown below. Comments shown in italics under some definitions are additional explanations that do not appear in the NFPA 70E standard.*

Arc-Flash Hazard: A dangerous condition associated with the possible release of energy caused by an electric arc.

Arc-Flash Hazard Analysis: A study investigating a worker's potential exposure to arc-flash energy, conducted for the purpose of injury prevention and the determination of safe work practices, arc-flash protection boundary, and the appropriate levels of PPE.

FPN No. 1: An arc-flash hazard may exist when energized electrical conductors or circuit parts are exposed or when they are within equipment in a guarded or enclosed condition, provided a person is interacting with the equipment in such a manner that could cause an electric arc. Under normal operating conditions, enclosed energized equipment that has been properly installed and maintained is not likely to pose an arc-flash hazard.

FPN No. 2: See Table 130.7(C)(9) for examples of activities that could pose an arc-flash hazard.

FPN No. 3: See 130.3 for arc-flash hazard analysis information.

Arc-Flash Protection Boundary: When an arc-flash hazard exists, an approach limit at a distance from exposed energized electrical conductor or circuit part within which a person could receive a second-degree burn if an electrical arc flash were to occur.

Comment: It is important to understand that staying outside the arc-flash protection boundary does not guarantee that a person will not be burned by an arc flash or arc blast. It means the person should receive a curable second-degree burn, rather than a more serious fatal one.

NFPA 70E DEFINITIONS

Arc-Flash Suit: A complete FR clothing and equipment system that covers the entire body, except for the hands and feet. This includes pants, jacket, and beekeeper-type hood fitted with a face shield.

Deenergized: Free from any electrical connection in a source of potential difference and from electrical charge; not having a potential different from that of the earth.

Comment: This is a key concept of NFPA 70E. *The safest way to work on electrical conductors and equipment is deenergized. See* Establishing an Electrically Safe Work Condition.

Disconnecting Means: A device, or group of devices, or other means by which the conductors of a circuit can be disconnected from their source of supply.

Comments:

- *Many different devices are permitted to serve as required disconnecting means for different types of equipment. These include wall switches and attachment plugs, under certain circumstances.*

- *The common field terms "disconnect switch" and "safety switch" are not used in* NFPA 70E *(or the* NEC*).*

- *Circuit breakers can be used as disconnecting means. Fuses by themselves are not considered disconnecting means. However, a fused disconnect switch or pullout block that simultaneously disconnects all ungrounded (phase) conductors of a circuit is considered a disconnecting means.*

Electrical Hazard: A dangerous condition such that contact or equipment failure can result in electric shock, arc-flash burn, thermal burn, or blast.

FPN: Class 2 power supplies, listed low-voltage lighting systems, and similar sources are examples of circuits or systems that are not considered an electrical hazard.

 *NFPA 70E* **DEFINITIONS**

Comment: NFPA 70E *safe work practices apply to systems energized at 50 volts and above, both AC and DC.*

Electrical Safety: Recognizing hazards associated with the use of electrical energy and taking precautions so that hazards do not cause injury or death.

Electrical Single-Line Diagram: A diagram that shows, by means of single lines and graphic symbols, the course of an electric circuit or system of circuits and the component devices or parts used in the circuit or system.

Electrically Safe Work Condition: A state in which an electrical conductor or circuit part to be worked on or near has been disconnected from energized parts, locked/tagged in accordance with established standards, tested to ensure the absence of voltage, and grounded if determined necessary.

Comment: This is a key concept of NFPA 70E. *The safest way to work on electrical conductors and equipment is deenergized. The process of turning off the electricity, verifying that it is off, and ensuring that it stays off while work is performed is called "establishing an electrically safe work condition." Many people call the process of ensuring that the current is removed "lockout–tagout"; however, lockout–tagout is only one step in the process.*

Employee in Charge: Although this term is not defined in Article 100, it is used a number places in the *NFPA 70E* standard. It indicates a team leader, foreman, general foreman, or other person responsible for directing electrical work and supervising electrical workers.

Energized: Electrically connected to or is a source of voltage.

Flame Resistant (FR): The property of a material whereby combustion is prevented, terminated, or inhibited following the application of a flaming or non-flaming source of ignition, with or without subsequent removal of the ignition source.

 *NFPA 70E* DEFINITIONS

FPN: Flame resistance can be an inherent property of a material, or it can be imparted by a specific treatment applied to the material.

Comment: The worker must understand that wearing FR clothing does not guarantee that a person will not be burned by an arc flash or arc blast. It means the person wearing FR clothing should receive a curable second-degree burn, rather than a more serious fatal one.

Incident Energy: The amount of energy impressed on a surface, a certain distance from the source, generated during an electrical arc event. One of the units used to measure incident energy is calories per centimeter squared (cal/cm^2).

Limited Approach Boundary: An approach limit at a distance from an exposed energized electrical conductor or circuit part within which a shock hazard exists.

Comment: This term describes a distance from exposed energized electrical conductors or circuit parts beyond which the risk of electric shock is considered low. Workers at least this far away from the equipment do not have to take special safety precautions. [See Prohibited Approach Boundary, Restricted Approach Boundary*, and* Working On (energized electrical conductor and circuit parts)]. *Approach boundaries are related only to shock exposure; they have nothing to do with arc-flash protection.*

Live Parts: Energized conductive components.

Premises Wiring (System): That interior and exterior wiring, including power, lighting, control, and signal circuit wiring, together with all their associated hardware, fittings, and wiring devices, both permanently and temporarily installed. This includes: (a) wiring from the service point or power source to the outlets; or (b) wiring from and including the power source to the outlets where there is no service point. Such wiring does not include wiring internal to appliances, luminaires, motors, controllers, motor control centers, and similar equipment.

NFPA 70E DEFINITIONS

Comment: As in the National Electrical Code, *this general term covers all power, communications, and control wiring of a building or similar structure from the service point, point of entry, or other source to the outlets.*

Prohibited Approach Boundary: An approach limit at a distance from an exposed energized electrical conductor or circuit part within which work is considered the same as making contact with the exposed energized electrical conductor or circuit part.

Comment: [See Prohibited Approach Boundary, Restricted Approach Boundary, *and* Working On (energized electrical conductor and circuit parts).] *Approach boundaries are related only to shock exposure; they have nothing to do with arc-flash protection.*

Qualified Person: One who has skills and knowledge related to the construction and operation of the electrical equipment and installations and has received safety training to recognize and avoid the hazards involved.

Comment: In general, both NFPA 70E *and the* National Electrical Code *define what tasks must be performed and what precautions must be taken to create a safe installation, but neither document specifies who shall perform them. Licensed electricians frequently perform electrical construction and maintenance work, but not all jurisdictions have electrician licensing. Engineers, technicians, specialized installers, and maintenance personnel also do electrical work, and many jurisdictions permit owners to perform electrical work on their own property. For these reasons,* NFPA 70E *uses the term "qualified person" in many places to indicate competency. The same term is used in the* NEC. (*See* Unqualified Person.)

Restricted Approach Boundary: An approach limit at a distance from an exposed energized electrical conductor or circuit part within which there is an increased risk of shock, due to electrical arc-over combined with inadvertent movement, for personnel working in close proximity to the exposed energized electrical conductor or circuit part.

 *NFPA 70E* DEFINITIONS

Comment: See Limited Approach Boundary, Prohibited Approach Boundary, *and* Working On (exposed energized electrical conductor or circuit part). *Approach boundaries are related only to shock exposure; they have nothing to do with arc-flash protection.*

Service Point: The point of connection between the facilities of the serving utility and the premises wiring.

Comment: This definition is a crucial concept in both NFPA 70E *and the* National Electrical Code. *Conductors and equipment on the customer side (load side—downstream) of the service point are covered by* NFPA 70E *and* NEC *rules. The* NEC *does not cover conductors and equipment on the utility side (line side—supply side—upstream) of the service point. Typically, these are constructed according to the* National Electrical Safety Code (NESC), *or the serving utility's own rules.*

Shock Hazard: A dangerous condition associated with the possible release of energy caused by contact or approach to energized electrical conductors, or circuit parts.

Step Potential: A ground potential gradient difference that can cause current flow from foot to foot through the body.

Comment: See Touch Potential.

Switch, Isolating: A switch intended for isolating an electric circuit from the source of power. It has no interrupting rating, and it is intended to be operated only after the circuit has been opened by some other means.

Touch Potential: A ground potential gradient difference that can cause current flow from hand to hand, hand to foot, or another path, other than foot to foot, through the body.

Comment: See Step Potential.

NFPA 70E DEFINITIONS

Unqualified Person: A person who is not a qualified person.

Comment: Unqualified persons are those not trained in the construction, operation, and safety aspects of electrical equipment and systems. See Qualified Person.

Working On (energized electrical conductor and circuit parts):
Coming in contact with an energized electrical conductor or circuit part with the hands, feet, or other body parts; with tools, probes, or test equipment, regardless of the PPE a person is wearing. There are two categories of "working on." *Diagnostic (testing)* is taking readings or measurements of electrical equipment with approved test equipment that does not require making any physical change to the equipment. *Repair* is any physical alteration of electrical equipment (such as making or tightening connections, removing or replacing components, and such).

Comment: See Prohibited Approach Boundary.

■ GENERAL PROTECTION FROM ELECTRICAL INJURIES

Most people think of PPE only as clothing or rubber insulating products. This text has already discussed PPE that guards against the hazards associated with shock and arc flash, even though some of these items might not normally be thought of as personal protective equipment. This section considers general PPE that can help prevent injuries against those and other hazards in electrical work. The following topics are not covered by *NFPA 70E* but are important for safe electrical practices:

• Hard hats

• Spectacles (safety glasses)

• Face shields and viewing windows

• Voltage-rated hand tools

• Safety grounds (clusters)

Hard Hats

Hard hats are intended to provide head protection from falling objects and bumping and to prevent the head from contacting energized conductors **(see Figure 44)**. Gravity usually causes falling objects to travel straight down from an elevated position. The forces of falling objects are applied directly to the top of the protective helmets and distributed onto the worker's head by the helmet's suspension system. Sometimes a falling object strikes a fixed object or structure and is directed into the worker's head from the side. The helmet assembly must distribute the kinetic energy contained in the falling object in this instance as well.

GENERAL PROTECTION FROM ELECTRICAL INJURIES

FIGURE 44   Protective helmets (hard hats). (Courtesy of Tasco-safety.com)

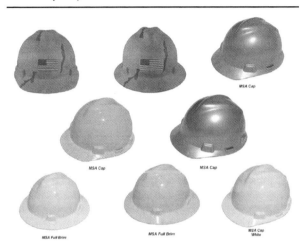

MSA Cap

MSA Cap

MSA Cap

MSA Full Brim

MSA Full Brim

MSA Cap White

The test setup for the different types of exposure to the hazards of falling objects or bumping must account for the various directions of the forces applied to the hard hat. National consensus standards use types to differentiate these exposures:

- Type I helmets are tested to distribute downward forces adequately.

- Type II helmets are tested to distribute both downward and lateral forces adequately.

Suspension systems may have four or six points of support for the hard hat **(see Table 8)**. Six-point systems provide greater distribution of the impact.

▟ GENERAL PROTECTION FROM ELECTRICAL INJURIES

TABLE 8   Classes of Hard Hats

Class	Maximum Voltage Rating	Equivalent Older Designation
E	20 kV	B
G	2.2 kV	A
C	No rating	C

NOTE: Class C hard hats are conductive.
Source: ANSI Z89

ANSI Z89.1 *Requirements*

One main ANSI national consensus standard now offers guidelines for head protection: *ANSI Z89.1, Requirements for Industrial Head Protection.*

ANSI does not write or publish standards. Instead, it selects and designates standards-developing organizations (SDOs) as secretariat organizations for standards. ANSI assigned the 2003 edition of the standards covering protective helmets to the International Safety Equipment Association (ISEA). Previously, the American Society of Safety Engineers (ASSE) had served as the secretariat for standards covering protective helmets.

- As a result of the change in secretariat, some structural changes were made to the affected documents. One change eliminated *ANSI Z89.2, Safety Requirements for Industrial Protective Helmets for Electrical Workers*, and integrated the requirements for protection from electrical shock into *ANSI Z89.1*, which is the standard that covers impact and penetration. Other changes modified the class designated for electrical shock protection from Classes A, B, and C to Classes E, G, and C.

GENERAL PROTECTION FROM ELECTRICAL INJURIES

- The change in secretariat for the national consensus standards occurred several years after applicable OSHA standards were issued. OSHA references to ANSI standards are to editions of the national consensus standards that were effective when the regulation was last promulgated. *ANSI Z89.2* was affected by the infrequent revision of OSHA standards. Although *ANSI Z89.2* has been eliminated, it is still referenced in *29 CFR 1910.6*. Therefore, users should ensure that they locate and use protective equipment that complies with the latest edition of the applicable national consensus standard.

- Where the hazard analysis indicates a chance for an elevated object to fall, employers are required to ensure that workers wear head-protective equipment. Basic requirements for protection from falling objects are defined in *ANSI Z89.1*. If the hazard analysis indicates a chance that the worker's head might be exposed to electrical shock as well as falling objects, current OSHA regulations indicate that helmets also must comply with *ANSI Z89.2-1971*. *ANSI Z89.2* is identified in *29 CFR 1910.6* as incorporated into the OSHA standards by reference.

OSHA Requirements

In 29 CFR 1910.132, OSHA requires employers to ensure that workers wear hard hats that provide protection from falling objects.

- Employers must execute and document a hazard analysis to decide what type of head protection workers should use.

- If the potential injury is limited to a worker bumping his or her head on an obstruction, the head protection need only consist of bump caps, which are lightweight and less restricting than heavier hard hats.

- However, use of bump caps is restricted to applications where falling objects are not anticipated.

GENERAL PROTECTION FROM ELECTRICAL INJURIES

A requirement for protective helmets that reduce the electrical shock hazard is cited in 29 CFR 1910.135(a)(2). In 1910.268(j)(1), Class B head protection is required when a worker might be exposed to high-voltage electrical contact. [Note that "Class B" was a designation in the now superseded *ANSI Z89.2*. The current *ANSI Z89.1* calls this head protection Class E.] In 1910.335(a)(1)(4), OSHA requires that workers wear head protection where danger from shock or burns is possible. The employer is responsible for ensuring that employees comply with these requirements.

ANSI Z89.1-2003 assigns helmets to a class as defined by voltage. Note that Class C hard hats are conductive and should not be worn by workers who are or may be exposed to an energized, uninsulated electrical conductor.

Selection and Use

- Hard hats are available in many different colors.

- Some employers provide workers in each discipline or craft with a unique color.

- In other instances, the color of the worker's hard hat indicates contract workers or specific contractors.

- In either instance, however, the hard hat should provide the impact protection and electrical shock protection defined in *ANSI Z89.1-2003*. The best alternative is for all workers to wear head protection rated as Class E protective hard hats.

- Hard hats are made from various moldable materials. Polyethylene is common, because it can be molded easily into the required form and offers excellent resistance to abrasion and breaking. Hard hats may be constructed from other materials, provided the completed assembly meets the impact, penetration, and insulating characteristics defined in *ANSI Z89.1*.

GENERAL PROTECTION FROM ELECTRICAL INJURIES

- Current national consensus standards for protective helmets define tests for impact, penetration, and electrical conductivity.

- However, ignition and flammability characteristics for protection from arc-flash events are not addressed.

- Although ignition and flammability are important for use by fire fighters, exposure to an arc-flash event is not currently considered important for construction of hard hats.

- A hard hat made from polyethylene and similar materials could ignite when exposed to an arcing fault and should be covered by a flame-resistant hood or similar product if the hazard/risk analysis indicates that the worker may be exposed to an arc-flash event.

- Workers sometimes put decals on the outer surface of their hard hat. If the hard hat is Class E or G (i.e., rated for use near exposed, energized electrical conductors), the decal should be nonconductive. Metal or conductive decals could be the cause of a short circuit and initiate an arcing fault.

- Chinstraps and other attachments are available for hard hats. Adding an attachment might provide additional utility; however, the attachment might increase the potential for damage from a different hazard. For instance, a chinstrap will prevent the helmet from falling off the head, but can increase the amount of fuel if it should be ignited in an arc-flash event.

Purchase specifications for hard hats must identify the characteristics of the desired helmet. It must indicate size (small, standard, or large), type (Type I or Type II), and color. When a hard hat is purchased, the suspension system must be purchased also. The suspension system must match the type and size of the hard hat. Incorrect application of the suspension system negates the protective characteristics of the protective helmet. Suspension systems and protective helmets from different manufacturers must not be intermingled. Hard hats are sold through distributors and local safety equipment suppliers.

GENERAL PROTECTION FROM ELECTRICAL INJURIES

Spectacles (Safety Glasses)

- National consensus standards define performance criteria, testing requirements, and required marking for safety glasses. *ANSI Z87.1, American National Standard for Occupational and Educational Eye and Face Protection Devices*, indicates that safety glasses can be either basic impact or high impact.

- Each level of impact protection requires different test setup and acceptance criteria. Purchase orders must specify if basic or high-impact safety glasses are required.

- The terms "eye glasses" and "safety glasses" are commonly used to mean an assembly of lenses together with a supporting frame that is worn to correct a person's vision or protect a person's eyes from impact **(see Figure 45)**.

- The term "spectacles" is used in consensus and regulatory standards to have the same meaning. Spectacles, however, excludes coverall safety glasses and goggles **(see Figure 46)**.

FIGURE 45   Safety glasses. (Courtesy of Ironwear)

FIGURE 46   Goggles. (Courtesy of Paulson Manufacturing)

In 29 CFR 1910.133(b)(2), OSHA suggests that protective equipment for eyes must comply with the requirements of *ANSI Z87.1*. This standard defines performance requirements that guide the construction of frames and lenses. By setting performance requirements, frames and support structures may be made from several different materials and with different construction, provided the overall assembly complies with the specified performance criteria.

Safety glasses, as such, are not electrical PPE. However, safety glasses could fall from a worker's face and initiate an arcing fault if the protective equipment is conductive. In some instances, the supporting frame for safety lenses is made of metal and is conductive. Unrestrained spectacles (safety glasses or otherwise) must not be worn when a worker is performing work on or near exposed live parts.

Face Shields and Viewing Windows

Every work task should begin with a hazard/risk analysis. Workers cannot select and wear adequate PPE until and unless all potential hazards have been identified. Unless workers are familiar with the nature and limits of protective characteristics for available PPE, the workers cannot choose the protective equipment with confidence. Workers, supervisors,

GENERAL PROTECTION FROM ELECTRICAL INJURIES

and managers must be trained to recognize the protective limits provided by equipment that meets national standards.

ANSI Z87.1 *Requirements*

Face shields can provide protection from several hazards.

- In some instances, a face shield is necessary to provide protection from impact. Grinding and cutting tasks usually generate flying objects that could injure a worker's face or eyes.

- A face shield that meets the impact and penetration requirements defined in *ANSI Z87.1, American National Standard for Occupational and Educational Eye and Face Protection Devices*, will provide adequate protection in this situation.

- When workers are required to work with and handle liquid chemicals, a face shield will provide protection from occasional splashes that might occur. If the face shield is impervious to attack from the chemical, the face shield will provide adequate protection. If the face shield used for protection from a chemical spill also meets the impact requirements of *ANSI Z87.1*, the same face shield may be worn for protection from both hazards.

- The same face shield that provides chemical and impact protection will filter some wavelengths of energy in the electromagnetic spectrum.

- Face shields that meet the requirements of *ANSI Z87.1* provide protection from nonionizing radiated energy. Some ultraviolet energy is filtered as well. However, face shields and other equipment covered by *ANSI Z87.1* do not provide protection from infrared energy or thermal energy **(see Figure 47)**.

An arcing fault generates significant thermal energy in the form of heated air and gasses that rush toward the worker's face at speeds approaching the speed of sound. Molten copper and parts and pieces

FIGURE 47   Arc-rated face shield. (Courtesy of Salisbury Electrical Safety LLC)

of the equipment might be rushing toward the worker's face at near the speed of sound. Significant electromagnetic energy is generated in an arcing fault and is radiated toward the worker's face at the speed of light. Workers should wear equipment that provides protection from all of these destructive components. An ordinary face shield intended for protection when using a grinder will not protect a worker from exposure to hazards associated with an arcing fault.

Direct contact between an exposed energized conductor or circuit part and a worker's face is unlikely. However, a worker's face is very likely to be exposed to the heated gasses and molten metal associated with an arcing fault.

- The worker's face and head must be protected from the destructive nature of these hazards.

- Therefore, the face shield must both meet impact requirements and provide adequate protection from the thermal hazard.

GENERAL PROTECTION FROM ELECTRICAL INJURIES

ASTM F2178-02 *Requirements*

ASTM F2178-02, Standard Test Method for Determining the Arc Rating of Face Protective Products, defines a testing method to establish a rating for face shields and viewing windows where the product will be exposed to an arcing fault.

- The rating system established in this standard suggests that products should be assigned an arc rating that essentially mirrors ratings for clothing defined in *ASTM F1506 Standard Performance Specification for Flame-Resistant Textile Materials for Wearing Apparel for Use by Electrical Workers Fxposed to Momentary Electric Arc and Related Thermal Hazard.*

- The arc rating should be based on the arc thermal performance value (ATPV) or the threshold break open energy (E_{BT}). Manufacturers assign arc ratings based on internal tests. No third-party testing process exists for establishing an arc rating.

- Face shield selection should be based on the expected available incident energy that is identified when the hazard/risk analysis is performed.

- A face shield that has a rating greater than the expected incident energy exposure should be selected.

- When face protection is an integral part of another item of protective apparel, the viewing window should have the same rating as the other protective apparel. Consult the manufacturer of the protective apparel.

- A face shield provides an edge on each side of the head of the person wearing it.

- A face shield obstructs the movement of the air and gasses rushing from the arcing fault during an exposure.

146

◉ GENERAL PROTECTION FROM ELECTRICAL INJURIES

- Objects such as molten metal will be stopped by the obstructing nature of the face shield. However, the obstructing nature and curved shape of a face shield also obstruct the movement of air and gasses rushing from an arcing fault. If the worker's head is positioned such that the rushing air is not directed at the front of the face shield, the curvature of the face shield could result in a vacuum being generated behind the face shield by the air and gasses rushing past. Lift is generated under an airplane wing by the same action **(see Figure 48)**.

All parts of a worker's body that are within the flash protection boundary must be protected from the effects of the potential arcing fault. If the worker's entire head is within the flash protection boundary, a face shield will not provide adequate protection. Only a switching hood with an adequately rated viewing window will provide the necessary protection in the case where the worker's head is within the flash protection boundary.

FIGURE 48   Illustration of lift.

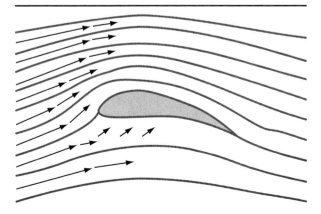

GENERAL PROTECTION FROM ELECTRICAL INJURIES

Voltage-Rated Hand Tools

In 29 CFR 1910.333(c)(2), OSHA suggests that voltage-rated tools must be used for all work where the hand tools might make contact with an exposed energized conductor. *NFPA 70E* also contains a similar requirement. *ASTM F1505, Standard Specification for Insulated and Insulating Hand Tools*, defines construction and testing of hand tools that are rated for use on circuits that are 1,000 Vac (volts of alternating current) or 1,500 Vdc (volts of direct current).

- The term *voltage-rated hand tools* refers to both insulated and insulating hand tools. Insulated hand tools are constructed from conductive material or components and have electrical insulation applied on the exterior surface.

- Insulating hand tools are constructed from nonconductive material. Insulating hand tools might have metal or other conductive inserts for reinforcement but, essentially, the tool must be constructed of nonconductive material **(see Figure 49)**.

FIGURE 49   Voltage-rated hand tools. (Courtesy of Salisbury Electrical Safety LLC)

⬛ GENERAL PROTECTION FROM ELECTRICAL INJURIES

- Voltage-rated hand tools are not intended to serve as primary protection from shock or electrocution.

- Although insulated hand tools might provide adequate shock protection for circuits below 1,000 Vac, workers must select and wear PPE that provides protection from shock and arc flash without considering the hand-tool rating.

- Although insulated and insulating hand tools include insulation from electrical sources of up to 1,000 Vac, the primary function of the insulation is to reduce the risk of initiating an arcing fault.

- The insulating coating of insulated hand tools may consist of a single layer, but normally it consists of two layers of contrasting colors.

- The interior layer provides 100 percent protection from shock to the full rating of the tool.

- The contrasting color of the exterior layer provides a method for inspecting the tool for damage.

- Any cut or abrasion that exposes any of the inner layers constitutes significant damage and suggests that the tool should not be used. The damaged tool should be replaced with a new tool.

- Workers should visually inspect each voltage-rated tool before each use. If the interior layer is visible, the tool should be discarded.

The manufacturer must mark voltage-rated hand tools as follows:

- Manufacturer's name or trademark

- Type or product reference

- Double triangle symbol **(see Figure 50)**

- 1,000-V

- Year of manufacture

GENERAL PROTECTION FROM ELECTRICAL INJURIES

FIGURE 50   Double triangle symbol.

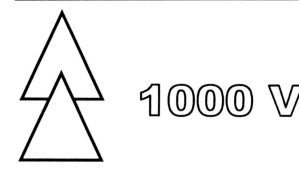

ASTM F1505 *Requirements*

ASTM F1505 defines the testing of voltage-rated (insulated and insulating) hand tools.

- Unless specifically approved for use at low temperatures, the tool may be used in ordinary atmospheric temperatures ranging from 20°C to 70°C.

- The standard requires the mechanical integrity of the tools to meet the same requirements as nonrated hand tools.

- Only hand tools that meet the requirements of *ASTM F1505* should be purchased.

- A storage toolbox should be purchased with the tools **(see Figure 51)**.

- The tools should be kept in the storage toolbox when not in use.

- The toolbox, with the tools, should be stored in a location that is clean and dry.

GENERAL PROTECTION FROM ELECTRICAL INJURIES

FIGURE 51   Sets of insulated hand tools.

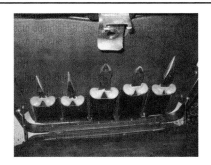

Safety Grounds (Clusters)

A deenergized electrical distribution circuit could be reenergized by several different means, which creates an unsafe condition.

- If the conductors are on cross arms, an energized conductor from a different circuit could fall onto one or more conductors of the deenergized circuit.

- If the circuit has equipment that is connected to multiple sources of energy, a second or third energy source could be operated to reenergize the deenergized circuit.

- The equipment could be back-fed through a transformer from a utilization circuit. In some instances, lockout–tagout could be in place and incorrectly implemented.

In each instance, a worker performing work on the distribution circuit could be electrocuted due to contact with the unintentionally energized conductor **(see Figure 52)**.

GENERAL PROTECTION FROM ELECTRICAL INJURIES

Where an exposure of this nature exists, the required electrically safe work condition does not exist. Workers must install safety grounds to control the potential exposure to shock and electrocution. When working on a deenergized conductor, the worker is likely to be a part of the deenergized circuit.

- To minimize exposure, the worker must install a device that will keep the potential difference between the conductor and surrounding conductive objects to a minimum.

- Installing an adequately rated conductor from the circuit being worked on and adjacent grounded surfaces will provide low impedance and thereby low potential difference.

- Installing a grounding cluster ensures that the upstream overcurrent device will see a short circuit and operate in a short time mode.

When current flows through an electrical conductor, a magnetic field is produced in the vicinity of the conductor.

GENERAL PROTECTION FROM ELECTRICAL INJURIES

- The strength of the field depends on the amount of current flow, and the direction of the magnetic field depends on the direction of the current flow in the conductor.

- The magnetic field associated with each conductor interacts with other magnetic fields and some other metallic components that might be nearby.

- When the amount of current flow is in the range of the ampacity rating of the conductor, the strength of the associated magnetic field produces little effect.

- However, when the amount of current approaches the available short-circuit range of a distribution circuit, the magnetic fields can become excessive.

- When the magnetic fields surrounding each conductor of a three-phase circuit interact with each other, substantial physical forces result. The construction of the ground cluster must be capable of conducting the maximum available fault current in a circuit.

The Right-Hand Rule

The right-hand rule predicts the direction of magnetic lines of force around an electrical wire that is conducting current. The right-hand rule says that if a person's right hand grips the electrical conductor so that the person's thumb is pointing in the direction of the current flow, the fingers on the hand illustrate the direction of the lines of magnetic force **(see Figure 53)**.

- As the amount of current increases, the strength of the magnetic lines of force also increases.

- When current is flowing in multiple conductors that are physically close to each other, the lines of magnetic force interact, resulting in strong physical forces applied to the conductors.

FIGURE 53   The right-hand rule.

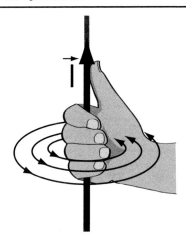

- If the conductors are a part of a ground cluster, the physical force tends to propel the conductor according to the interaction of the lines of magnetic force.

Performance Requirements

ASTM F855-04, Standard Specifications for Temporary Protective Grounds to Be Used on Deenergized Electric Power Lines and Equipment, defines performance requirements for components of these ground clusters and for the complete assembly. Although the possibility exists for employees to assemble adequate protective grounds, the adequacy of the overall construction is not dependable without performing tests defined in the standard. Manufacturers perform tests as a routine

GENERAL PROTECTION FROM ELECTRICAL INJURIES

part of the manufacturing process and assign fault duty ratings to the completed ground clusters. Only adequately rated ground clusters should be used **(see Figure 54)**; they are available through electrical distributors.

Temporary protective grounds should be installed as close to the work site as possible.

- A ground cluster should be selected that has a fault duty rating at least as great as the available fault current at the point of the work.

FIGURE 54 Ground clusters with four conductors. (Courtesy of Salisbury Electrical Safety LLC)

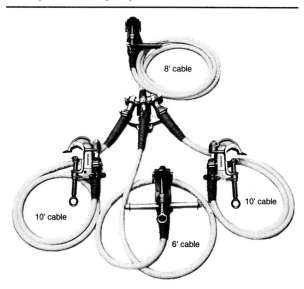

GENERAL PROTECTION FROM ELECTRICAL INJURIES

- Ground clusters must be marked to indicate the rating assigned by the manufacturer.

- After determining the necessary fault duty rating and selecting a ground cluster, the conductors, clamps, and connecting points should be visually inspected to ensure that the components of the ground cluster have not been damaged.

- If any sign of damage to the conductors or clamps is found, another ground cluster should be selected.

- Installing temporary ground clusters is one step required to establish an electrically safe work condition.

- Until the ground cluster has been satisfactorily installed, the worker should consider the conductor to be energized.

- Wearing appropriate FR protection, the worker should install the ground cluster with a live-line tool or while wearing voltage-rated protective equipment.

- The first connection should be to an adequately sized grounding conductor. Subsequent connections should be made to each phase conductor.

- Ground clusters should be removed in reverse order. Remove the connection to the grounded conductor after all connections to a phase conductor are removed.

- OSHA requires a zone of equipotential be established for exposed overhead conductors. A zone of equipotential means that all metal components, including conductors, are grounded in such a way that the worker is unlikely to reach outside the zone of protection **(see Figure 55)**.

- If the overhead line receives a lightning discharge at another location, the safety grounds might protect the worker, regardless of the location of the strike.

GENERAL PROTECTION FROM ELECTRICAL INJURIES

- Ground clusters that have been repaired or modified must be tested to ensure that the repaired equipment will pass the standard 30-cycle or 15-cycle voltage-drop values permitted by *ASTM F855*.

- Ground clusters should be subjected to the 3-cycle or 15-cycle voltage-drop tests defined in the standard on a regular basis as determined by conditions of use. However, the test interval must not exceed 3 years.

FIGURE 55   Illustration of zone equipotential.

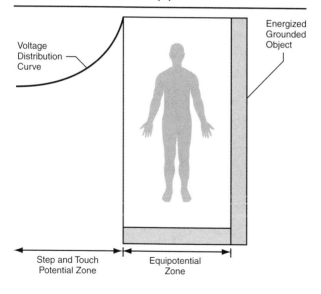

 **FIRST AID**

In addition to consistently wearing PPE, the best protection from the consequences of severe injury is the knowledge of First Aid practices and the ability to act in an emergency situation. The Emergency Care and Safety Institute is an educational organization created for the purpose of delivering the highest quality training to lay people and professionals in the areas of First Aid, CPR, AED, Bloodborne Pathogens, and related safety and health areas. The content of the training materials used by the Emergency Care and Safety Institute is approved by the American Academy of Orthopaedic Surgeons (AAOS) and the American College of Emergency Physicians (ACEP), two of the most respected names in injury, illness, and emergency medical care.

Visit www.ECSInstitute.org for more information.

Scene Survey

When approaching the scene of an emergency, do a 10-second scene survey to assess the following:

- Danger to the rescuer and to the victim. Scan the area for immediate dangers to yourself or to the victim. If the scene is unsafe, make it safe. If you are unable to do so, do not enter.

- Cause of injury or nature of illness. This helps to identify what is wrong.

- Number of victims. Determine how many people are involved. There may be more than one person, so look around and ask about others.

If there are two or more victims, first check those who are not moving or talking. These are the individuals who may need your help first.

 **FIRST AID**

How to Call for Help

To receive emergency assistance of every kind in most communities, simply call 9-1-1. At some government installations and industrial sites, an additional system may apply. This should be an element of a job briefing. In any case, be prepared to tell the Emergency Medical Services (EMS) dispatcher the following:

- Your name and phone number

- Exact location or address of emergency

- What happened

- Number of people

- Victim's condition and what is being done for the victim

Do not hang up until the dispatcher hangs up—the EMS dispatcher may be able to tell you how to care for the victim until the ambulance arrives.

Bleeding

1. Cover the entire wound with clean, dry cloth or sterile dressing.

2. Press against the wound for 5–10 minutes. If the bleeding does not slow or stop, press harder over a wider area. If the bleeding is from an arm or a leg, raise the limb above the heart level unless the arm or the leg is broken.

3. For a shallow wound, wash it with soap and water and flush with forceful, running water. For a deep wound, do not use soap and water under pressure. Deep wounds require cleaning by a medically trained person. Cover large, gaping wounds with sterile gauze pads.

4. When the bleeding stops or subsides, secure gauze or cloth snugly with a bandage. For a shallow wound, antibiotic ointment can be applied before the dressings. Do not use antibiotic ointment on a deep wound.

 **FIRST AID**

Protect yourself against diseases carried by blood by wearing disposable medical exam gloves, using several layers of cloth or gauze pads, using waterproof material such as plastic, or having the victim apply pressure using his or her own hand.

Ingested Poisons

Recognizing Ingested Poisoning

The signs of ingested poison include the following:

- Abdominal pain and cramping
- Nausea or vomiting
- Diarrhea
- Burns, odor, or stains around and in the mouth
- Drowsiness or unresponsiveness
- Poison container nearby

Care for Ingested Poison

1. Remove any objects from the victim's mouth.
   - Try to determine what poison was swallowed and how much.
   - Call the poison control center immediately for instructions: 800-222-1222.

2. Keep the victim on his or her left side to delay the poison from emptying into the small intestine where it gets into the bloodstream faster.

If the victim complains about burning sensations or if you see burns around the mouth, immediately give the victim milk or water.

 **FIRST AID**

Shock

Recognizing Shock

The signs of shock include the following:

- Altered mental status (agitation, anxiety, restlessness, and confusion)

- Pale, cold, and clammy skin, lips, and nail beds

- Nausea and vomiting

- Rapid breathing

- Unresponsiveness (when shock is severe)

Care for Shock

Even if there are no signs of shock, you should still treat seriously injured or suddenly ill victims for shock.

1. Place the victim on his or her back.

2. Raise the legs 6 to 12 inches (if spinal injury is not suspected). Raising the legs allows the blood to drain from the legs back to the heart.

3. Place blankets under and over the victim to keep the victim warm.

Burns

Care for Burns

1. Stop the burning! Use water or smother flames.

2. Cool the burn. Apply cool water or cool, wet cloths until pain decreases (usually within 10–40 minutes). Do not apply cold to more than an area equivalent to the size of the victim's entire chest or back (about 20% of the body surface area).

 **FIRST AID**

3. Apply aloe-vera gel on first-degree burns (skin turns red). Apply antibiotic ointment on second-degree burns (skin blisters). Apply non-stick dressing on second- and third-degree burns (full thickness; skin dies).

Seek medical attention if any of these conditions exist:

- Breathing difficulty

- Head, hands, feet, or genitals involved

- Victim's age is under 5 years or over 60 years

- Involves electricity or chemical exposure

- Second-degree burns cover more than an area equivalent to size of victim's entire back or chest

- Any third-degree burns

Electrical Burns

- Check the scene for electrical hazards.

- Check breathing and pulse; CPR may be needed.

Frostbite

Recognizing Frostbite

The signs of frostbite include the following:

- White, waxy-looking skin

- Skin feels cold and numb (pain at first, followed by numbness)

- Blisters, which may appear after re-warming

FIRST AID

Care for Frostbite

1. Move the victim to a warm place.

2. Remove tight clothing or jewelry from the injured part.

3. Place dry dressings between the toes and fingers.

4. Seek medical care.

Hypothermia

Recognizing Hypothermia

The signs of hypothermia include the following:

- Uncontrollable shivering

- Confusion, sluggishness

- Cold skin (even under clothing)

Care for Hypothermia

1. Get the victim out of the cold.

2. Prevent heat loss by:
    - replacing wet clothing with dry clothing.
    - covering the victim's head.
    - placing insulation (such as blankets, towels, coats) beneath and over the victim.

3. Have the victim lie down.

4. If the victim is alert and able to swallow, give him or her warm, sugary beverages.

5. Seek medical care for severe hypothermia (rigid muscles, cold skin on abdomen, confusion, or lethargy).

 **FIRST AID**

Heat Cramps

Recognizing Heat Cramps

The signs of heat cramps include the following:

- Painful muscle spasms during or after physical activity

Care for Heat Cramps

1. Have the victim stop activity and rest in a cool area.

2. Stretch the cramped muscle.

3. Remove any excess or tight clothing.

4. If the victim is responsive and not nauseated, provide water or a commercial sport drink (such as Gatorade® or Powerade®).

Heat Exhaustion

Recognizing Heat Exhaustion

The signs of heat exhaustion can include the following:

- Heavy sweating

- Severe thirst

- Weakness

- Headache

- Nausea and vomiting

Care for Heat Exhaustion

1. Have the victim stop activity and rest in a cool area.

2. Remove any excess or tight clothing.

FIRST AID

3. If the victim is responsive and not nauseated, provide water or a commercial sport drink (such as Gatorade® or Powerade®).

4. Have the victim lie down and raise his or her legs 6 to 12 inches.

5. Cool the victim by applying cool, wet towels to the victim's head and body.

Seek medical care if the condition does not improve within 30 minutes.

Heatstroke

Recognizing Heatstroke

The signs of heatstroke can include the following:

- Extremely hot skin

- Dry skin (may be wet at first)

- Confusion

- Seizures

- Unresponsiveness

Care for Heatstroke

1. Have the victim stop activity and rest in a cool area.

2. Call 9-1-1.

3. If the victim is unresponsive, open his or her airway, check breathing, and provide appropriate care.

4. Rapidly cool the victim by whatever means possible: cool, wet towels or sheets to the head and body accompanied by fanning, and/or cold packs against the armpits, the sides of the neck, and the groin.

 **FIRST AID**

Airway Obstruction

Management of Choking Responsive Victim

1. Check the victim for choking by asking, "Are you choking?" A choking person is unable to breathe, talk, cry, or cough.

2. Have someone call 9-1-1.

3. Position yourself behind the victim and locate the victim's navel.

4. Place a fist with the thumb side against the victim's abdomen just above the navel, grasp it with the other hand, and press it into the victim's abdomen with quick inward and upward thrusts. Continue thrusts until the object is removed or the victim becomes unresponsive.

If the victim becomes unresponsive, call 9-1-1 and give CPR. Each time you open the airway to give a breath, look for an object in the mouth or throat and, if seen, remove it.

Adult Cardiopulmonary Resuscitation (CPR)

1. Check responsiveness by tapping the victim and asking, "Are you okay?" If the victim is unresponsive, roll the victim onto his or her back.

2. Have someone call 9-1-1, and have someone else retrieve an AED if available.

3. Open the airway using the head tilt–chin lift method (lift the chin with one hand and tilt the head back with the other hand).

4. Check for breathing for 5 to 10 seconds by looking for chest rise and fall and listening and feeling for breathing. If the victim is breathing, place him or her in the recovery position. If the victim is not breathing, go to the next step.

⎑ FIRST AID

5. Give 2 rescue breaths (1 second each), making the chest rise.
   Pinch the victim's nose, place your mouth over the victim's mouth,
   and give 2 breaths (1 second each), pausing between each breath.
   If the first breath does not make the chest rise, re-tilt the head and
   try the breath again and then proceed to the next step. If both
   breaths make the chest rise, go to the next step.

6. Perform CPR.
   - Place the heel of one hand on the center of the chest between the
     nipples. Place the other hand on top of the first hand.
   - Depress the chest 1.5 to 2 inches.
   - Give 30 chest compressions at a rate of about 100 per minute,
     allowing the chest to return to its normal position after each
     compression.
   - Open the airway, and give 2 breaths (1 second each).

7. Continue cycles of 30 chest compressions and 2 breaths until an
   AED is available, the victim starts to move, EMS takes over, or you
   are too tired to continue.

Heart Attack

Recognizing a Heart Attack

Prompt medical care at the onset of a heart attack is vital to survival
and the quality of recovery. This is sometimes easier said than done
because many victims deny they are experiencing something as serious
as a heart attack. The signs of a heart attack including the following:

- Uncomfortable pressure, fullness, squeezing, or pain in the center of
  the chest lasting 2 minutes or longer. It may come and go.

- Pain may spread to either shoulder, the neck, the lower jaw, or either
  arm.

- Any or all of the following: weakness, dizziness, sweating, nausea, or
  shortness of breath.

 **FIRST AID**

Care for a Heart Attack

1. Seek medical care by calling 9-1-1. Medications to dissolve a clot are available but must be given early.

2. Help the victim into the most comfortable resting position, usually sitting with legs up and bent at the knees. Loosen clothing. Be calm and reassuring.

3. If the victim is alert, able to swallow, and not allergic to aspirin, give one adult aspirin or two to four chewable children's aspirin.

4. If the victim has been prescribed medication for heart disease, such as nitroglycerin, help the victim to use it.

5. Monitor the victim's breathing.

Stroke

Recognizing a Stroke

- Sudden weakness or numbness of the face, an arm, or a leg on one side of the body

- Blurred or decreased vision, especially on one side of the visual field

- Problems speaking

- Dizziness or loss of balance

- Sudden, severe headache

![] FIRST AID

Care for a Stroke

1. Call 9-1-1.

2. If the victim is responsive, lay the victim on his or her back with the head and shoulders slightly elevated.

3. If the victim is unresponsive, open the airway, check breathing, and provide care accordingly. If the unresponsive victim is breathing, place the victim on his or her side to keep the airway clear.

First Aid, Rescue, and CPR

NFPA 70E requires employees to have first aid and emergency training but does not provide details. OSHA regulations require that at least one person on each job site be trained in first aid and CPR.

Many electrical contractors require foremen and general foremen to have first aid training. First aid and CPR training are available from many sources, including the following:

• Emergency Care and Safety Institute, www.ECSInstitute.org

• Local fire and rescue departments

• Community colleges

Many industrial plants, other facilities, and construction projects maintain a nurse on site.

REFERENCES

ANSI/ASSE Z87.1, American National Standard for Occupational and Educational Eye and Face Protection Devices. New York, NY: American National Standards Institute/American Society of Safety Engineers, 1998.

ANSI/ISA S82.02.01, Electrical and Electronic Test, Measuring, Controlling, and Related Equipment, General Requirement. New York, NY: American National Standards Institute/Instrument Society of America, 1999.

ANSI Z87.1, Practice for Occupational and Educational Eye and Face Protection. New York: American National Standards Institute, 1996.

ANSI Z535.4, Product Safety Signs and Labels. New York, NY: American National Standards Association, 1998.

ANSI Z89.1, Requirements for Industrial Head Protection. New York, NY: American National Standards Institute, 2003.

ANSI Z89.2, Safety Requirements for Industrial Protective Helmets for Electrical Workers. (Superseded by *ANSI Z89.1* and no longer published, although it remains a reference in *29 CFR 1910.6.*)

ASTM D120, Standard Specification for Rubber Insulating Gloves. Conshohocken, PA: American Society of Testing and Materials, 2002.

ASTM D178, Standard Specification for Rubber Insulating Matting. Conshohocken, PA: American Society of Testing and Materials, 2002.

ASTM D1048, Standard Specification for Rubber Insulating Blankets. Conshohocken, PA: American Society of Testing and Materials, 2002.

ASTM D1051, Standard Specification for Rubber Insulating Sleeves. Conshohocken, PA: American Society of Testing and Materials, 2002.

REFERENCES

ASTM D6413, Standard Test Method for Flame Resistance of Textiles. Conshohocken, PA: American Society of Testing and Materials, 1999.

ASTM F479, Specification for In-Service Care of Insulating Blankets. Conshohocken, PA: American Society of Testing and Materials, 2002.

ASTM F496, Specification for In-Service Care of Insulating Gloves and Sleeves. Conshohocken, PA: American Society of Testing and Materials, 2002.

ASTM F711, Standard Specification for Fiberglass-Reinforced Plastic (FRP) Rod and Tube Used in Live-Line Tools. Conshohocken, PA: American Society of Testing and Materials, 2002.

ASTM F1116, Standard Test Method for Determining Dielectric Strength of Dielectric Footwear. Conshohocken, PA: American Society of Testing and Materials, 2005.

ASTM F2413, Standard Specification for Performance Requirements for Foot Protection. Conshohocken, PA: American Society of Testing and Materials, 2005.

ASTM F1449, Standard Guide for Care and Maintenance of Flame-Resistant Clothing. Conshohocken, PA: American Society of Testing and Materials, 2001.

ASTM F1505, Standard Specification for Insulated and Insulating Hand Tools. Conshohocken, PA: American Society of Testing and Materials, 2001.

ASTM F1506, Standard Performance Specification for Flame-Resistant Textile Materials for Wearing Apparel for Use by Electrical Workers Exposed to Momentary Electric Arc and Related Thermal Hazards. Conshohocken, PA: American Society of Testing and Materials, 2002.

REFERENCES

ASTM F1959, Standard Test Method for Determining the Arc Rating of Materials for Clothing. Conshohocken, PA: American Society of Testing and Materials, 2005.

ASTM F855-04, Standard Specifications for Temporary Protective Grounds to Be Used on De-energized Electric Power Lines and Equipment. Conshohocken, PA: American Society of Testing and Materials, 2004.

ASTM F2178-02, Standard Test Method for Determining the Arc Rating of Face Protective Products. Conshohocken, PA: American Society of Testing and Materials, 2002.

Doughty, Richard L., et al. "Predicting Incident Energy to Better Manage the Electric Arc Hazard on 600V Power Distribution Systems." PCIC Paper PCIC-98-36. Paper presented at the Forty-Fifth Annual Conference of the IAS/IEEE Petroleum and Chemical Industry Committee, Indianapolis, Indiana, September 28–30, 1998.

Doughty, Richard L., et al., "Testing Update on Protective Clothing and Equipment for Electric Arc Exposure." PCIC Paper PCIC-97-35. Paper presented at the Forty-Fourth Annual Conference of the IAS/IEEE Petroleum and Chemical Industry Committee, Banff, Alberta, September 15–17, 1997.

IEC 61010, Safety Requirements for Electrical Equipment for Measurement, Control and Laboratory Use. Geneva, Switzerland: International Electrotechnical Commission, 2002.

IEEE Standard 978, Guide for In-Service Maintenance and Electrical Testing of Live-Line Tools. Piscataway, NJ: Institute of Electrical and Electronics Engineers, 1984.

IEEE Standard 1584, IEEE Guide for Performing Arc-Flash Hazard Calculations. New York, NY: Institute of Electrical and Electronics Engineers, 2002.

Jones, Ray A. and Jane G. Jones, The Electrical Safety Program Book. Quincy, MA/Sudbury, MA: NFPA/Jones and Bartlett, 2003.

REFERENCES

Lee, Ralph H. "Pressures Developed by Arcs." *IEEE Transactions on Industry Applications*, Vol. 1A-23, No. 4, July/August 1987. Piscataway, NJ: IEEE, 1987.

The National Electrical Code (ANSI/NFPA 70). Quincy, MA: National Fire Protection Association, 2005.

The National Electrical Safety Code (ANSI/IEEE C2). New York, NY: Institute of Electrical and Electronics Engineers, 2007.

NFPA 70E, Standard for Electrical Safety Requirements for Employee Workplaces. Quincy, MA: National Fire Protection Association, 2009.

UL 1244, Standard for Electrical and Electronic Measuring and Testing Equipment. Requirements. Northbrook, IL: Underwriters Laboratories, 2000.

U.S. Department of Labor. Occupational Safety and Health Administration. OSHA Regulations 29 CFR 1910.300-399, Subpart S, "Electrical." Washington, DC.

U.S. Department of Labor. Occupational Safety and Health Administration. OSHA Regulations 29 CFR 1910.132-139, Subpart I, "Personal Protective Equipment." Washington, DC.

U.S. Department of Labor. Occupational Safety and Health Administration. OSHA Regulations 29 CFR 1910.137, "Electrical Protective Equipment." Washington, DC.

U.S. Department of Labor. Occupational Safety and Health Administration. OSHA Regulations 29 CFR 1910.269, "Electric Power Generation, Transmission, and Distribution." Washington, DC.

2009 CALENDAR

JANUARY

S	M	T	W	T	F	S
				1	2	3
4	5	6	7	8	9	10
11	12	13	14	15	16	17
18	19	20	21	22	23	24
25	26	27	28	29	30	31

FEBRUARY

S	M	T	W	T	F	S
1	2	3	4	5	6	7
8	9	10	11	12	13	14
15	16	17	18	19	20	21
22	23	24	25	26	27	28

MARCH

S	M	T	W	T	F	S
1	2	3	4	5	6	7
8	9	10	11	12	13	14
15	16	17	18	19	20	21
22	23	24	25	26	27	28
29	30	31				

APRIL

S	M	T	W	T	F	S
			1	2	3	4
5	6	7	8	9	10	11
12	13	14	15	16	17	18
19	20	21	22	23	24	25
26	27	28	29	30		

MAY

S	M	T	W	T	F	S
					1	2
3	4	5	6	7	8	9
10	11	12	13	14	15	16
17	18	19	20	21	22	23
24	25	26	27	28	29	30
31						

JUNE

S	M	T	W	T	F	S
	1	2	3	4	5	6
7	8	9	10	11	12	13
14	15	16	17	18	19	20
21	22	23	24	25	26	27
28	29	30				

JULY

S	M	T	W	T	F	S
			1	2	3	4
5	6	7	8	9	10	11
12	13	14	15	16	17	18
19	20	21	22	23	24	25
26	27	28	29	30	31	

AUGUST

S	M	T	W	T	F	S
						1
2	3	4	5	6	7	8
9	10	11	12	13	14	15
16	17	18	19	20	21	22
23	24	25	26	27	28	29
30	31					

SEPTEMBER

S	M	T	W	T	F	S
		1	2	3	4	5
6	7	8	9	10	11	12
13	14	15	16	17	18	19
20	21	22	23	24	25	26
27	28	29	30			

OCTOBER

S	M	T	W	T	F	S
				1	2	3
4	5	6	7	8	9	10
11	12	13	14	15	16	17
18	19	20	21	22	23	24
25	26	27	28	29	30	31

NOVEMBER

S	M	T	W	T	F	S
1	2	3	4	5	6	7
8	9	10	11	12	13	14
15	16	17	18	19	20	21
22	23	24	25	26	27	28
29	30					

DECEMBER

S	M	T	W	T	F	S
		1	2	3	4	5
6	7	8	9	10	11	12
13	14	15	16	17	18	19
20	21	22	23	24	25	26
27	28	29	30	31		

2010 CALENDAR

JANUARY

S	M	T	W	T	F	S
					1	2
3	4	5	6	7	8	9
10	11	12	13	14	15	16
17	18	19	20	21	22	23
24	25	26	27	28	29	30
31						

FEBRUARY

S	M	T	W	T	F	S
	1	2	3	4	5	6
7	8	9	10	11	12	13
14	15	16	17	18	19	20
21	22	23	24	25	26	27
28						

MARCH

S	M	T	W	T	F	S
	1	2	3	4	5	6
7	8	9	10	11	12	13
14	15	16	17	18	19	20
21	22	23	24	25	26	27
28	29	30	31			

APRIL

S	M	T	W	T	F	S
				1	2	3
4	5	6	7	8	9	10
11	12	13	14	15	16	17
18	19	20	21	22	23	24
25	26	27	28	29	30	

MAY

S	M	T	W	T	F	S
						1
2	3	4	5	6	7	8
9	10	11	12	13	14	15
16	17	18	19	20	21	22
23	24	25	26	27	28	29
30	31					

JUNE

S	M	T	W	T	F	S
		1	2	3	4	5
6	7	8	9	10	11	12
13	14	15	16	17	18	19
20	21	22	23	24	25	26
27	28	29	30			

JULY

S	M	T	W	T	F	S
				1	2	3
4	5	6	7	8	9	10
11	12	13	14	15	16	17
18	19	20	21	22	23	24
25	26	27	28	29	30	31

AUGUST

S	M	T	W	T	F	S
1	2	3	4	5	6	7
8	9	10	11	12	13	14
15	16	17	18	19	20	21
22	23	24	25	26	27	28
29	30	31				

SEPTEMBER

S	M	T	W	T	F	S
			1	2	3	4
5	6	7	8	9	10	11
12	13	14	15	16	17	18
19	20	21	22	23	24	25
26	27	28	29	30		

OCTOBER

S	M	T	W	T	F	S
					1	2
3	4	5	6	7	8	9
10	11	12	13	14	15	16
17	18	19	20	21	22	23
24	25	26	27	28	29	30
31						

NOVEMBER

S	M	T	W	T	F	S
	1	2	3	4	5	6
7	8	9	10	11	12	13
14	15	16	17	18	19	20
21	22	23	24	25	26	27
28	29	30				

DECEMBER

S	M	T	W	T	F	S
			1	2	3	4
5	6	7	8	9	10	11
12	13	14	15	16	17	18
19	20	21	22	23	24	25
26	27	28	29	30	31	

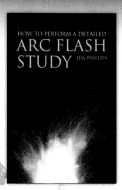